Springer Theses

Recognizing Outstanding Ph.D. Research

Aims and Scope

The series "Springer Theses" brings together a selection of the very best Ph.D. theses from around the world and across the physical sciences. Nominated and endorsed by two recognized specialists, each published volume has been selected for its scientific excellence and the high impact of its contents for the pertinent field of research. For greater accessibility to non-specialists, the published versions include an extended introduction, as well as a foreword by the student's supervisor explaining the special relevance of the work for the field. As a whole, the series will provide a valuable resource both for newcomers to the research fields described, and for other scientists seeking detailed background information on special questions. Finally, it provides an accredited documentation of the valuable contributions made by today's younger generation of scientists.

Theses are accepted into the series by invited nomination only and must fulfill all of the following criteria

- They must be written in good English.
- The topic should fall within the confines of Chemistry, Physics, Earth Sciences, Engineering and related interdisciplinary fields such as Materials, Nanoscience, Chemical Engineering, Complex Systems and Biophysics.
- The work reported in the thesis must represent a significant scientific advance.
- If the thesis includes previously published material, permission to reproduce this must be gained from the respective copyright holder.
- They must have been examined and passed during the 12 months prior to nomination.
- Each thesis should include a foreword by the supervisor outlining the significance of its content.
- The theses should have a clearly defined structure including an introduction accessible to scientists not expert in that particular field.

More information about this series at http://www.springer.com/series/8790

Jian-Wei Liu

Well-Organized Inorganic Nanowire Films

Assemblies and Functionalities

Doctoral Thesis accepted by
University of Science and Technology
of China, Hefei, China

Springer

Author
Dr. Jian-Wei Liu
Department of Chemistry
University of Science and Technology of China
Hefei, Anhui
China

Supervisor
Prof. Shu-Hong Yu
Department of Chemistry
University of Science and Technology of China
Hefei
China

ISSN 2190-5053 ISSN 2190-5061 (electronic)
Springer Theses
ISBN 978-981-10-3946-1 ISBN 978-981-10-3947-8 (eBook)
DOI 10.1007/978-981-10-3947-8

Library of Congress Control Number: 2017932019

Printed on acid-free paper

This Springer imprint is published by Springer Nature
The registered company is Springer Nature Singapore Pte Ltd.
The registered company address is: 152 Beach Road, #21-01/04 Gateway East, Singapore 189721, Singapore

Supervisor's Foreword

In the past two decades, accompanied by the development of nanoscience and nanotechnology, the research field of nanowire films has flourished with research activity both fundamental and applied. Ordered nanowire films of a diverse range of compositions and architectures were assembled after synthesis showing tailored physical and chemical properties. Macroscopic-scale integrating nanowires from disordered nanowire solution into ordered or controlled assembled structures will provide an advanced understanding of self-assembly and a new way for tailoring the properties of nanowires for application potential.

In this thesis, in the introduction section nanowire assembly strategies are summarized and many related literatures are also highlighted. However, we still meet serious issues and encounter severe challenges in fabrication of macroscopic nanowire assemblies.

Using simple solution-based strategies, such as microwave-assisted method and solvothermal process, uniform and high-quality one-dimensional nanowires were synthesized with sufficient characterization. The rational design and precise shape and size control of one-dimensional nanostructures through solution-based approaches plays an important role in fabricating functional nanodevices. Macroscopic well-defined periodic Tellurium and $W_{18}O_{49}$ nanowire mesostructures can be produced by the Langmuir–Blodgett (LB) technique without any extra hydrophobic pretreatment or functionalities. Based on the reactivity of the Te nanowire arrays, Te-telluride heteronanowire films were obtained with component and structure controllable. Using LB technique, Ag nanowires and Te nanowires were selected to construct co-assemblies. Besides LB technique, liquid–liquid interface, electron beam, and flowing shear interaction can provide driving forces for nanowire assembly. The ability to manipulate the assembly of nanowires, together with an understanding of the structure-dependent properties will offer a great opportunity to rationally design new nanodevices with substantially enhanced performance.

This thesis provides fresh insight into the nanowire assembles and their functionalities. Based on these insights, versatile interface-induced nanowire assembly strategies are innovatively developed to optimize the applications. These findings also open opportunities for designing new nanodevices with ordered nanowire films.

December 2016

Prof. Shu-Hong Yu
Hefei, China

Parts of this thesis have been published in the following journal articles:

Liu, J. W.; Chen, F.; Zhang, M.; Qi, H.; Zhang, C. L.; Yu, S. H., Rapid Microwave-Assisted Synthesis of Uniform Ultra long Te Nanowires, Optical Property, and Chemical Stability. *Langmuir* **2010,** *26* (13), 11372–11377.

Liu, J. W.; Zhu, J. H.; Zhang, C. L.; Liang, H. W.; Yu, S. H., Mesostructured Assemblies of Ultrathin Superlong Tellurium Nanowires and Their Photoconductivity. *J. Am. Chem. Soc.* **2010,** *132* (26), 8945–8952.

Liu, J. W.; Zhang, S. Y.; Qi, H.; Wen, W. C.; Yu, S. H., A general strategy for self-assembly of nanosized building blocks on liquid/liquid interfaces. *Small* **2012,** *8* (15), 2412–20.

Liu, J.-W.; Liang, H.-W.; Yu, S.-H., Macroscopic-Scale Assembled Nanowire Thin Films and Their Functionalities. *Chem. Rev.* **2012,** *112* (8), 4770–4799.

Liu, J.-W.; Wang, J.-L.; Huang, W.-R.; Yu, L.; Ren, X.-F.; Wen, W.-C.; Yu, S.-H., Ordering Ag nanowire arrays by a glass capillary: A portable, reusable and durable SERS substrate. *Sci. Rep.* **2012,** *2*, 987.

Liu, J.-W.; Xu, J.; Liang, H.-W.; Wang, K.; Yu, S.-H., Macroscale Ordered Ultrathin Telluride Nanowire Films, and Tellurium/Telluride Hetero-Nanowire Films. *Angew. Chem. Int. Ed.* **2012,** *51* (30), 7420–7425.

Liu, J.-W.; Xu, J.; Ni, Y.; Fan, F.-J.; Zhang, C.-L.; Yu, S.-H., A Family of Carbon-Based Nanocomposite Tubular Structures Created by in Situ Electron Beam Irradiation. *ACS Nano* **2012,** *6* (5), 4500–4507.

Liu, J. W.; Zheng, J.; Wang, J. L.; Xu, J.; Li, H. H.; Yu, S. H., Ultrathin W18O49 Nanowire Assemblies for Electrochromic Devices. *Nano Lett.* **2013,** *13* (8), 3589–93.

Liu, J.-W.; Huang, W.-R.; Gong, M.; Zhang, M.; Wang, J.-L.; Zheng, J.; Yu, S.-H., Flexible Electronics: Ultrathin Hetero-Nanowire-Based Flexible Electronics with Tunable Conductivity. *Adv. Mater.* **2013,** *25* (41), 5909–5909.

Liu, J. W.; Wang, J. L.; Wang, Z. H.; Huang, W. R.; Yu, S. H., Manipulating Nanowire Assembly for Flexible Transparent Electrodes. *Angew. Chem. Int. Ed.* **2014,** *53* (49), 13477–13482.

Liu, J.-W.; Xu, J.; Hu, W.; Yang, J.-L.; Yu, S.-H., First Systematic Synthesis of Tellurium Nanostructures and Their Optical Property: From Nanoparticles, to Nanorods, Nanowires, and Nanotubes. *Chem Nano Mat* **2016**, 392–393.

Acknowledgements

I offer my deepest and sincerest gratitude first and foremost to my respected supervisor Prof. Shu-Hong Yu. I feel so lucky to be one of his Ph.D. students and receive his guidance through the nearly six-year study and all the stages of this thesis. What he taught me is not only how to do the research work itself but also how to become a respectable man with well-disciplined, dedicated, and hard-working attitude, as well as the exploration spirit. Without his consistent encouragement and illuminating instruction, I would still grope in the darkness, amid thunder and lightning. During my six-year Ph.D. study, he provided me with advanced facilities and shiny thinking which I would cherish for my whole life. He also gave me numerous opportunities to touch the edge of science and let me attend many academic conferences to broaden my scientific horizons. He loves all of his students and treats us as his own children. He feels very happy whenever we make a small success and always encourages us whenever we meet difficulties. He is very hard-working and spends every minute on scientific adventure. It is his influence and encouragement that makes me decide to become a scholar and a teacher. One could not wish for a better supervisor than mine, Prof. Shu-Hong Yu.

Thanks a lot for all the brothers and sisters in Yu laboratory for being with me, and I will always be there for you. It would not have been possible to have such a wonderful Ph.D. life without you. I also greatly thank my parents, wife, brother, and other relatives. It is your selfless love that gives me courage to follow my dream.

Acknowledgements

Contents

Chapter 1
Introduction

1.1 Background and Significance

A nanowire is a structure with a diameter of the order of a nanometer. When the diameter puts the radial dimension of the nanowire at or below certain characteristic lengths, such as the Bohr radius, the wavelength of the light, and phonon mean-free path, quantum mechanical effects become important [1, 2]. With a large surface-to-volume ratio and two-dimensional confinement, nanowires show unique optical, magnetic, and electronic properties [3–9]. Moreover, the large aspect ratio of nanowires, as an ideal energy transport material, can direct the conduction of quantum particles such as electrons, and photons improving their technological application. Significant advances have been made in last years for the fabrication of a wide range of low-dimensional nanostructured materials with good qualities and in large quantities [5, 10–19]. A reasonable assembly strategy is needed not only to create complex structures with new collective properties, but also to make full use of ordered nanowire architectures to fabricate advanced functional devices. Nanoscale assembly using linear nanostructures has become one of the most active areas of research in materials science [20–22]. The controllable synthesis of nanowires has achieved a lot of results, which provides the feasibility for the development of related properties. In order to integrate these nanowires as assembling elements on a macroscopic scale, it is necessary to develop an appropriate method to assemble them. This level of control also allows for a deeper understanding of the assembly phenomenon which has long been a subject of research on small-sized nanowire fabrication. At the same time, these technologies also make up for many different approaches in nanosynthesis. Over the past two decades, the self-assembly patterns in chemistry, physics, and biology have matured scientifically to a level of sophistication that one can begin to explore their potential in nanosynthesis. The process of spontaneously transmuting an existing nanomaterial into an ordered, organized, or specific functional system is referred to self-assembly (SA). The direct or indirect interaction of the nanomaterial with its environment is considered responsible for the self-assembly [23, 24]. As an important

J. Liu, *Well-Organized Inorganic Nanowire Films*, Springer Theses,
DOI 10.1007/978-981-10-3947-8_1

process of changing the disordered structure into an ordered structure, self-assembly can be characterized by the minimum free energy of the system. At the same time, self-assembly that is either through a direct or indirect correlation is typically related with the thermodynamic equilibrium. Nanowire and nanowire assemblies have unique electrical, optical, magnetic, and catalytic properties, not only due to a significant increase in surface area-to-volume ratio but also due to the presence of collective nanosize characteristics due to cross-line alignment. In any case, the ability to produce nanowire meshes, arrays, and composites depends on our understanding and control of the assembly process of these materials. Because of the large aspect ratio of the nanowires, they are often bent and kinked, hindering them from forming ordered nanostructures in self-assembly process [25, 26]. In the early twenty-first century, many impressive reviews about the nanowires were published, mainly about the synthesis and characterization of one-dimensional nanowires [1, 27]. With the transfer of research interest to the assembly of nanowires, the nanowire assembly field is undergoing rapid expansion and becomes one of the most active research fields in nanoscience. However, there are only a few review articles that describe the development of this research field. Recently, Yang's group studied the Langmuir–Blodgett (LB) method [28, 29] for the assembly of nanocrystals and nanowires and published some influential review articles. As a flexible and versatile approach, the LB assembly method can also be used for functional organic nanowires, reviewed by Ariga and his collaborators. At the same time, the Lieber team also reviewed the synthesis and application of nanowires on the basis of their excellent work. For the assembly of functional organic nanowires, Ariga and Hu's group reviewed their controllable assembly and growth [26, 30]. From the perspective of the interaction force, the Gates group published a short review of nanowire assembly [31].

Here, we intend to demonstrate the process in the emerging field of nanowires. In general, this chapter briefly describes the (i) synthesis of nanowires with controlled morphology; (ii) fabrication of nanowires at wafer level through the interface, including LB technology, liquid–liquid interface; (iii) assembly of nanowires by in situ electron-beam irradiation; and (iv) fabrication of nanodevices based on well-defined nanowire films. (Te) nanowires with a diameter of 20 nm and lengths of several tens of micrometers can be rapidly synthesized by the microwave-assisted method. In addition to the microwave-assisted method, a systematic study on the synthesis of nanostructures is conducted from sub-nanocrystals, nanorods, and nanowires to nanotubes through a simple solution.

1.2 Thin Films Composed of Assembled Ordered Nanowire Arrays

Nanowires have very interesting applications, but their disordered structure makes their use in device fabrication (e.g., microelectronics, optoelectronics) very problematic because these devices often require the arrangement of highly ordered structures [32–35]. Reasonable assembly strategies are required to construct complex structures with novel collective properties and to fabricate nanoscale components in practical dimensions. It is sometimes necessary to use a wire-based film for device assembly. To achieve the fundamental application of nanowires (NWs) in these or other nanotechnology fields, we need to have a controllable and predictable assembly of highly ordered structures. Moreover, the next generation of high-performance electrical, optoelectronic, and electromagnetic systems can be realized with thin films consisting of one-dimensional nanostructures with controlled position, orientation, and spacing of various lengths [36]. Complex nanowire assembly structures require not only the influence of individual wires but also the control of the connections between the nanowires. Furthermore, it is necessary to develop suitable assembly methods to produce films consisting of highly ordered nanostructures, in order to achieve low-cost applications of one-dimensional nanomaterials.

A number of review articles [29, 30, 37, 38] for nanowire assembly have been published. However, how to manipulate nanowire arrays or individual nanowire in accordance with a controllable manner still remains challenging. This lack of capability is now hampering the integration of devices and integrated circuits at the macroscale. In general, there are two methods for assembly of nanowires: a "top-down" method based mainly on photolithography and a typical self-assembly method, which is "bottom-up" process. The self-assembly method has a lower production cost and higher production efficiency, in comparison with the prior strategies and also provides more flexibility for the selection of the functional material and the synthesis process. However, much effort is required to further develop this approach. Here, we provide a summary of the "bottom-up" assembly of one-dimensional nanostructures using different methods and mechanisms (Fig. 1.1). This research has attracted the interest of physicists, chemists, and materials scientists, especially in trying to discover "green" methods for synthesizing one-dimensional nanostructures. The integration of these nanowires requires the development of suitable assembly methods. These techniques can compensate for many different routes in nanosynthesis.

1.2.1 Interface-Induced Nanowire Assembly

1.2.1.1 Assembly of Nanowires by the Langmuir–Blodgett Technique

Langmuir–Blodgett (LB) method is a traditional method for the synthesis of the monolayer at the gas–liquid interface. Today, LB technology has proven to be a high-throughput, low-cost, easy-to-integrate approach for assembling nanoscale structural elements into tightly packed nanosuperstructures and low-density well-defined patterns [39, 40]. Figure 1.2a shows the LB assembly process for orienting the nanowires. The nanomaterials are first dispersed in an immiscible and volatile nonpolar solvent and diffused to a water-supported interface using a

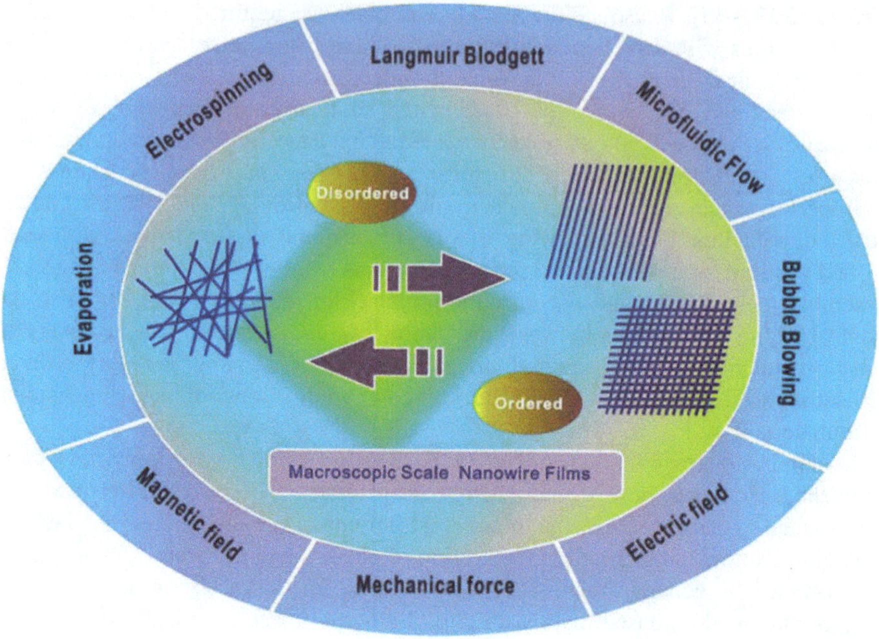

Fig. 1.1 Schematic illustration of macroscopic-scale nanowire thin film assembling methods that make disordered one-dimensional nanowire ordered. Reproduced with permission from Ref. [6] Copyright 2012 American Chemical Society

microadjustment syringe. With the evaporation of the solvent, when the surface pressure is controlled, the sample will be slowly compressed; at the same time, Langmuir thin film can be prepared by a single layer of nanostructures [29]. After the monolayer film is prepared at the interface, the monolayer film can be dynamically controlled by compression or expansion during the LB assembly process, thereby obtaining a monolayer film having a suitable structure, on the basis of the degree of control of the compression. These monolayer films can be repetitively deposited by a technique of vertical immersion (Langmuir–Blodgett) or horizontal lifting (Langmuir–Schaefer) to realize the structure of the multilayer film on a solid substrate. This provides a method for preparing thin film devices of a particular thickness and provides additional possibilities for fundamental scientific research on the relationship between assembled layers. Yang's group pioneered a research project in the area of nanoparticle–nanowire assembly using LB technology and recently published some influential review articles [41, 42]. Next, this universal approach has attracted a great deal of attention worldwide. The first is the assembly of the functional Ge nanowires, the Ge nanowires are dispersed in the organic solvent, and the tightly aligned monolayer is easily obtained by the LB process (Fig. 1.2b). In addition, ordered nanowire films, such as ordered VO_2 nanowires (Fig. 1.2c), Ag nanowires (Fig. 1.2d), and PbS nanowires (Fig. 1.2e, f), have been prepared [43, 44].

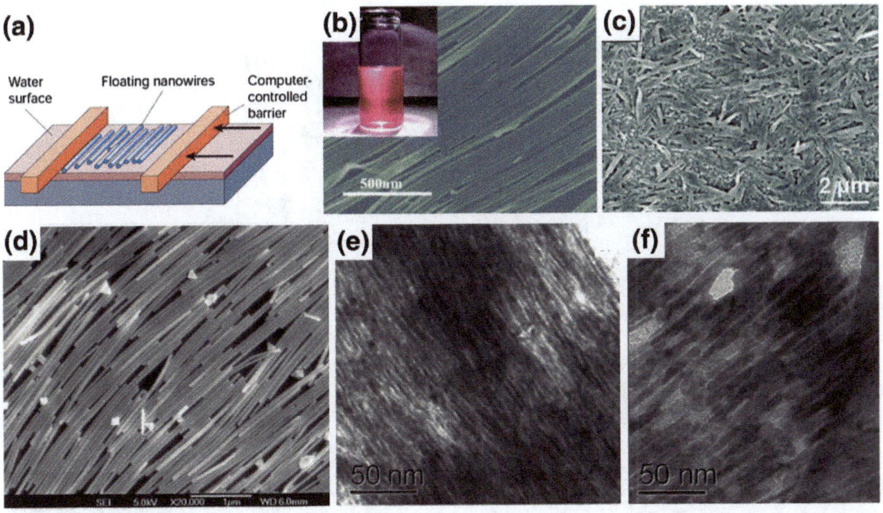

Fig. 1.2 **a** Illustration of LB nanowire assembly process. Reproduced with permission from Ref. [23] Copyright 2003 Nature Publishing Group. **b** Scanning electron microscope (SEM) image of the aligned Ge NW film. *Inset* photograph of a Ge nanowire suspension in chloroform. Reproduced with permission from Ref. [43] Copyright 2005 American Chemical Society. **c** Typical SEM image of VO_2 nanowire LB films deposited at surface pressures of 39.5 mN/m. Reproduced with permission from Ref. [6] Copyright 2009 American Chemical Society. **d** Scanning electron microscopy image of the silver nanowire monolayer deposited on a silicon wafer. Reproduced with permission from Ref. [42] Copyright 2003 American Chemical Society. **e** TEM image of large-scale assembly of PbS wires at surface pressures of 23 mN/m and **f** Transmission electron microscopy (TEM) image of thicker PbS nanowires after coalescence at surface pressures of 25 mN/m for 90 min. **e–f** Reproduced with permission from Ref. [44] Copyright 2007 American Chemical Society

Although LB technology has proved to be a good general method for assembling one-dimensional nanostructures, it still has some inherent limitations. For example, in the case of hydrophilic nanoscale units, it is necessary to functionalize the surface of the nanomaterials by hydrophobic ligand in the LB experiments, which greatly limits their future applications [28]. Recently, Yu group has fabricated a clear periodic ordered structure [40] of hydrophilic Te nanowires with an aspect ratio of over 10^4 without any pretreatment of hydrophobic pretreatment. In order to avoid hydrophobization, the hydrophilic superfine Te nanowires were dissolved in amphiphilic dimethylformamide (DMF) before being added to the air–water interface and then dispersed to the mixture of DMF and chloroform. Figure 1.3 presents nanowires assembly of ultrathin Te nanowires and the small-angle X-ray scattering pattern (SAXRD) measurements of assembled nanowire, which indicates that the period of the single-layer Te nanowires is 7.16 nm. Assembled monolayer nanowires make it possible to construct nanomesh-like mesostructures or complex multilayered nanowire structures on a planar substrate that can precisely control the included angle (Fig. 1.3d–f). In addition to the Te nanowires, this improved LB technique has also been found to be used to assemble other flexible, one-dimensional, ultralong nanostructures, such as Ag_2Te nanowires and Pt nanotubes [43].

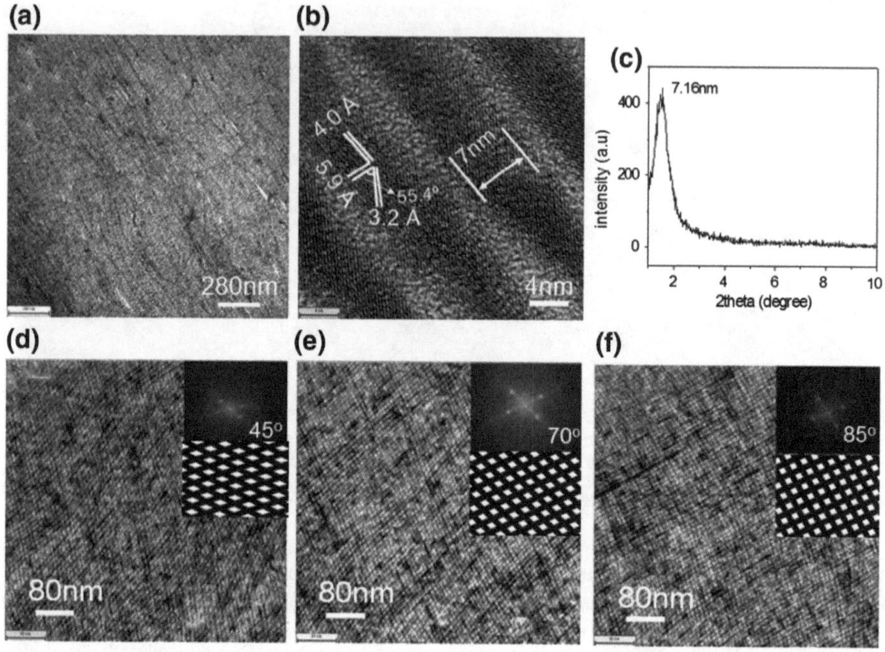

Fig. 1.3 **a**, **b** TEM and High-resolution transmission electron microscopy (HRTEM) images of the monolayer assembly of Te nanowires. **c** SAXRD pattern measured on the monolayer of aligned Te nanowires. **d–f** Crossed reproduced with permission from Ref. [40] Copyright 2010 American Chemical Society

1.2.1.2 Evaporation-Induced Assembly of Nanowires

The kinetics of the evaporation process is very complex and comprises many challenging problems in soft-matter physics [45–48], whereas the assembly of nanowires facilitated by evaporation is very common [49–53]. When some nanomaterials are dispersed in a volatile solvent, a drop of such solution is allowed to dry on the solid surface or the liquid–gas interface, the material in the liquid will first take a random orientation, and eventually, the dual function of the material and the substrate (material–material and material–substrate interactions) leaves a dense and neat assembly [54]. The final morphology and uniformity of the material vary widely, depending on the size distribution of the nanomaterial and the degree of dewetting of the solvent during the assembly process. The nanomaterials initially diffuse throughout the droplet solution, and the nanostructures will gradually build up as the solvent evaporates. The final outward can bring all the dispersed material to the boundary. During the drying process, the liquid evaporated from the boundary will be replaced by an internal liquid, and the contact line of the droplet experience stick-slip motion. In the competition between the fixed and the capillary force, a large displacement is produced by a small displacement, leaving a small number of nanostructures randomly distributed on the substrate [54]. When the contact line is fixed, the liquid evaporating from the boundary is replaced by the internal liquid. Because the outward flow of solvent brings the dispersed nanostructures to the boundary, the nanostructures are more concentrated on the boundary than the center. When the degree of aggregation of the nanostructures on the boundaries increases to a critical level, the nanostructures start to change from homogeneous to liquid crystalline. The formed liquid crystalline state finally precipitates at the substrate at the edge, which helps to secure the contact line [55]. As more nanostructures are brought to the edge due to the outward flow, the liquid crystal phase gradually expands. As the contact angle decreases to a certain value, the contact line begins to recede until the solvent evaporates completely [55]. In this section, some recently reported nanowire assembly by evaporation is presented as an example [56]. Yan and his coworkers reported on the synthesis of uniform nanowires with diameters of about 1.8 nm and lengths of several microns. The ultrafine nanowires can be spontaneously assembled into a three-dimensional superstructure by the parallel arrangement after the evaporation of the mixed solution of cyclohexane and ethanol [57]. Assembly of the nanowires facilitated by evaporation can be controlled by confined spaces, resulting in a series of ordered nanostructures [58–61]. Wang et al. reported a simple method for preparing large-scale ordered coaxial organic nanowires by simply evaporating the solvent in a confined space [62]. Figure 1.4a shows the spherically flattened evaporation setup throughout the evaporation process and how the evaporation direction of the solvent is controlled in a fume hood at room temperature. When the solution is allowed to dry on the substrate, a number of ordered circular nanowires are coaxially aligned and form concentric rings on both sides (Fig. 1.4b is a schematic). The confined space consists of a glass slide and a spherical lens made of fused silica. A drop of dimethyl quinacridone (DMQA), selected as a nonvolatile solute, is dropped into space between the slide and the lens. At the same time, a steady stream of nitrogen up-flow is added around the device to aid in the removal of the solvent. When the solution is dried in the film, it leaves numerous ordered rings of nanowires that assemble and form concentric rings on both surfaces [62]. The SEM images in Fig. 1.4c, d show the formation of coaxial DMQA nanowire loops during solvent evaporation. The length, density, and periodicity of a nanostructure array can be controlled by controlling the evaporation rate. Coaxial rings with pronounced periodicity can extend up to several hundred microns, and each ring has a series of DMQA nanowires [63]. The Luis M. Liz-Marzan's team in Spain reviewed the process of nanoparticles self-assembly by using templates or adding external fields to change the distribution of energy or entropy [64]. More recently, their team reported the binary self-assembly of Au nanowires containing Au nanoparticles and Au nanorods, obtained by evaporation of the solvent (Fig. 1.4e–g). Long nanowires can direct the assembly of nanoparticles and nanorods to form an extended ordered array. In addition, nanowires can modulate the distance between the nanoparticles and nanorods to alter the overall optical response of the film [65]. Choi and coworkers have reported self-assembly of nanowires by evaporation of nanowire colloidal droplets on superhydrophobic surfaces. This self-assembly is achieved by the interaction of the nanowires with the micrometer tip structure of the

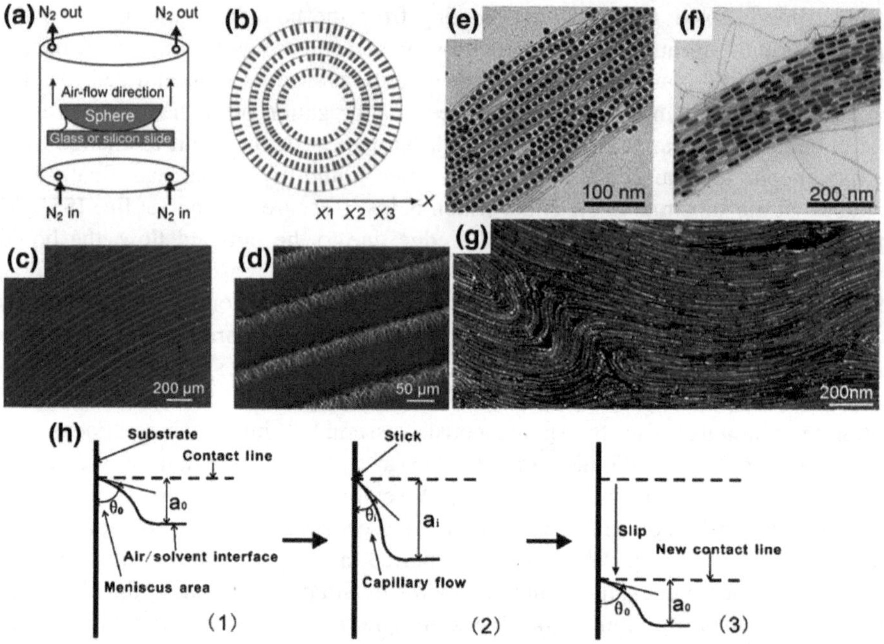

Fig. 1.4 **a** Schematic representation of a sphere-planar evaporation apparatus. **b** Shows the concentric rings of the prepared organic nanowires. **c–d** SEM images of concentric rings of DMQA nanowires formed during solvent evaporation, with different magnifications. Reproduced with permission from Ref. [63] Copyright 2011 WILEY-VCH. **e–g** A TEM image of a binary assembly comprising Au nanoparticles **e** and Au nanorods **f** at a constant nanowire concentration. **g** A SEM image of a binary assembly of Au nanoparticles induced by a flexible array of Au nanowires. Reproduced with permission from Ref. [72] Copyright 2010 WILEY-VCH. **h** Stepwise schematically show the contact line movement process on a vertical substrate induced by solvent evaporation. In the initial stage *1*, the contact angle between the solvent and the substrate is a_i and the meniscus height is a_0. In stage *2,* as the evaporation time elapses, the contact angle decreases, and the meniscus height increases, since the contact line adheres to the substrate. In stage *3,* when the meniscus extends to its maximum, the contact line cannot stick, but slide to a new location. Therefore, the contact angle and meniscus height are returned to their initial values, respectively. Reproduced with permission from Ref. [49] Copyright 2008 WILEY-VCH

superhydrophobic surface [66]. Xu and his co-authors describe a surface solvent evaporation method for the synthesis of radial anthracene nanowires and their coaxial self-assembled circular arrays [67].

Welland and co-authors reported a self-assembly method [68] for obtaining individual films from amyloid fibrils by evaporation. The films obtained are very ordered and highly rigid, with Young's modulus as high as 5–7 GPa, which is comparable to the highest value of the protein material in nature. Briefly, the film is obtained by transferring 1 ml of hydrogel-containing nanowires onto a flat polytetrafluoroethylene film. After 24 h, the solvent and the volatile acid are allowed to evaporate completely and the resulting protein film can be transferred with tweezers [69]. The film comprising the plasticizer can be cut into a preferred shape with a blade while the film obtained from pure protein fibers are more brittle and may be broken into pieces by the use of tweezers. Solvent losses because of evaporation during drying can result in capillary flow and solute transport [70, 71]. The substrate is treated to acquire hydrophilic surface so that the solvent has a nonzero contact angle with the surface and the contact line is fixed due to the roughness of the surface (Fig. 1.4h). Normally, the contact line is fixed in its initial position [72]. The physical phenomena of solvent evaporation and contact line pinning force the solutes (nano building blocks) in the droplet to flow toward the contact line (i.e., capillary flow) [72]. Since the contact action has anisotropic nature, the tip of the nanowire is guided with the flow of the solution, demonstrating clearly the self-assembly of nanorods in the direction of dewetting direction. In general, the combined effect of capillary flow and strong interaction guides the formation of ordered nanowire arrays [55]. Conversely, the fixation and dewetting of the contact line lead to competition between the pinning and the depinning forces. During the solvent evaporation, the contact angle decreases ($\theta_i < \theta_0$) and the meniscus height increases ($a_i > a_0$). These changes increase the depinning force without changing the pinning force. As the depinning force increases above a pinning force, the contact line becomes unstable and slides to a new position, thereby creating a new nanowire band. The deposited nanowires cause an increase in the surface roughness and the fixing force. Larger nanowire densities also increase the pinning force, so that the meniscus can be strained longer before the contact line slides. As a result, the spacing of the nanowire arrays is expected to become larger as the evaporation rate is higher [40].

1.2.1.3 Other Nanowire Assemblies at Complex Interfaces

The interface offers a unique heterogeneous dielectric environment when the phases with completely different dielectric constants are in contact. In recent years, interface assembly has been proved to be an important platform for the assembly of nanosized monolayers, based on the interfacial-ordering effects [73]. There are two main methods for generating ordered films at the interface: The first is to exchange the ligands [62] on the nanowires before forming the interface by mixing the suspensions in another solvent, the second is to form the interface by adding inorganic salts, organic solvents, or ultrasound to change the interaction between nanowires [64]. Choi et al. described a nanowire self-assembly method, demonstrating formation of three-phase (liquid–solid–gas) contact line with the tip of the structure by a well-defined superhydrophobic structure on the surface of a template and, when the colloidal droplet of nanowires retreats in evaporation [74], it leads to self-assembly of the nanowire [65]. Lee and their collaborators have developed a "one-step" process [66] for generating and assembling monocrystalline organic nanowires at organic solvent–water interfaces. Solvent evaporation shows a vital role in the assembly process. When dichloromethane (DCM) evaporates, the organic nanowires are trapped at the DCM–water interface due to compressive forces. The density of the nanowires increases on further evaporation of DCM, resulting in the shrinking of DCM interface. When the solvent is completely evaporated, a compressed monolayer of ordered nanowires appears on the surface of the water. In this process, the disordered nanowires are compressed into ordered arrays. These films can be shifted directly to any substrate, or stacked one layer at a time to a multilayer film for device application. The elements team reported the synthesis and assembly of a long, up to 1-mm column of dyed porphyrin molecules at a solid–liquid interface, highly ordered in space and in parallel. First, 3 ml of a chloroform solution containing a porphyrin dye at a concentration of 4.8×10^{-6} M was released onto the mica sheet. After evaporation, a highly ordered pattern of equidistant, nearly parallel, filamentary structures in a large area (up to about 3 mm^2) can be observed. By combining the self-assembly with the dewetting process, the orientation process occurs at the same time that the droplets of the porphyrin evaporate on the surface [75]. Recently, Yu group introduced a novel and effective method for the fabrication of independent ordered Ag-nanowire films by three-phase interfacial assembly [76]. Disordered Ag nanowires before assembly and ordered Ag nanowires after assembly are presented in Fig. 1.5a, b. Because the Ag nanowires are highly ordered, optical films exhibit a luster of thin films similar to silver mirrors. Using this simple method, Ag nanowires with large aspect ratios can be arranged closely and in parallel with each other. The self-assembly process takes place by dispersion of proper amount of Ag-nanowire (dispersed in water) droplets onto the surface of chloroform to form a water–oil–gas three-phase interface. For the aggregation and self-assembly of long Ag nanowires, the three-phase interface is necessary. The schematic representation of three-phase self-assembly of nanowires is presented in Fig. 1.5c–e, illustrating the formation of an Ag-nanowire array obtained by a three-phase interface method [76]. Figure 1.5f–i shows the photographs corresponding to the steps in Fig. 1.5e. Initially, the oil–water–gas three-phase interface is obtained by dropping an appropriate amount of Ag-nanowire aqueous solution on the surface of chloroform. First, the Ag nanowires are shifted from the water–oil interface to the water–air interface step by step through the oil–water–gas boundary because of the evaporation of chloroform. Next, the Ag nanowires begin to self-assemble at the contact line of the water phase and the wall of the beaker that is represented as the water–air–substrate interface. Finally, the Ag-nanowire film on the water–air–substrate interface starts to grow into continuous and highly ordered nanowire film. Thus, the three-phase interface plays an important role in the assembly process and can drive the movement of silver nanowires and self-assembly [69]. In addition, many factors such as the PVP molecules, the evaporation of the oil phase and the capillary force are considered key aspects to promote the self-assembly of the long silver nanowires. Silver nanowires are prepared by reduction of $AgNO_3$ in the presence of PVP by a polyol process which has good dispersion in water due to the protective layer of PVP on its surface. PVP as a surfactant offers stronger interactions among the neighboring Ag nanowires, which may improve the nanowires binding ability and align parallel to each other. The washing of PVP-coated Ag nanowires results in a decline in self-assembly order of silver nanowires [76]. Recently, Yu and coworker have reported a method of building large-scale self-assembled nanofilms at room temperature in a water–oil interface, which is composed of nanocubes, nanoparticles, nanowires, and nanosheets [77]. A series of liquid–liquid interface systems can be selected as a platform for

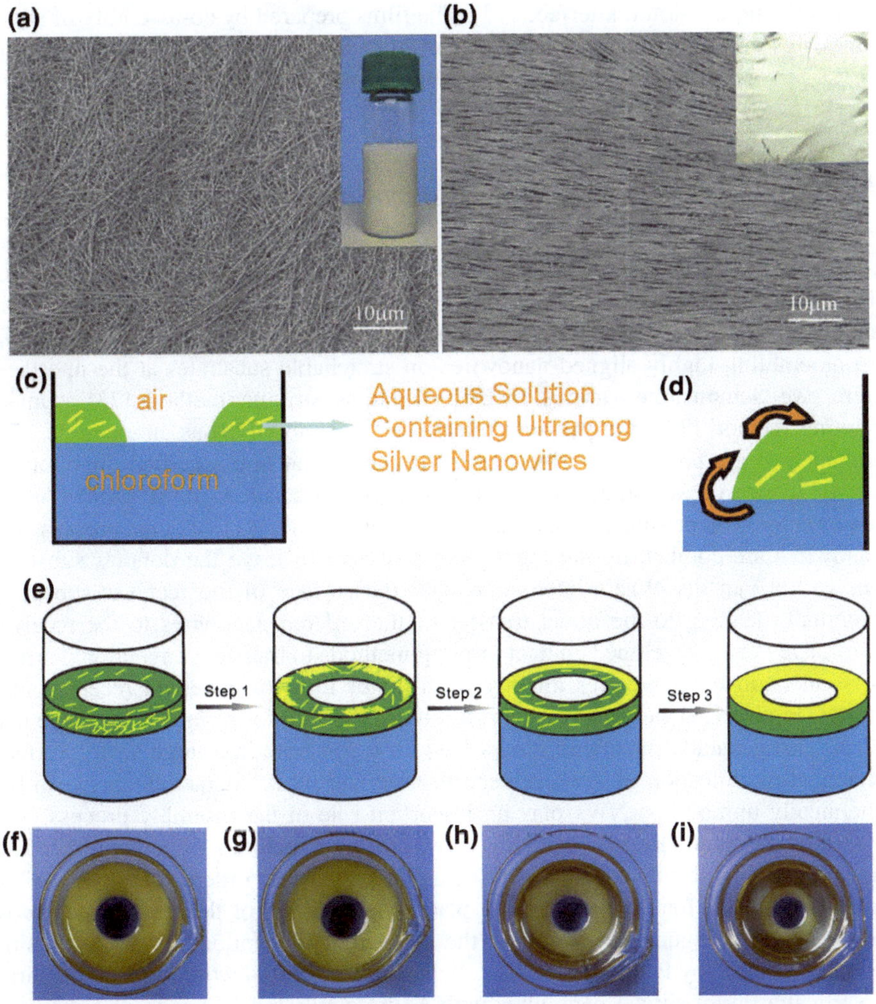

Fig. 1.5 **a** SEM image of disordered Ag nanowires. **b** SEM image of assembled nanowires. Inset of **a** and **b** are optical photographs of Ag nanowires dispersed in distilled water, and the assembled nanowires are transferred to a silicon wafer. **c** A cross section of the three-phase interface. **d** A schematic representation of the movement of the Ag nanowires as the oil phase evaporates in the first step. **e** A schematic view of the stage in which the Ag nanowires form a thin film at the three-phase interface. **f–i** The optical images of the assembly steps illustrated in Scheme **e**. Reproduced with permission from Ref. [69] Copyright 2010 WILEY-VCH

the construction of nanofilms, which also deepens the understanding of nanoscale aggregation and self-assembly. The flat and uneven substrate can be used to transfer nanofilms, and the thickness of the nanofilm can be controlled by layer-by-layer deposition. Furthermore, nanoparticles and nanowires can be co-assembled very well at the liquid–liquid interface [77]. The films prepared by co-assembly of silver nanowires and silver nanoparticles exhibit high SERS intensities, and the specific SERS performance is described later in this chapter.

1.2.2 Nanowire Assemblies by Mechanical Force

Uniform and controlled assembly of nanowire materials with high order offer one of the significant bottleneck tasks facing the integration of nanowires for electronic applications. Mechanical force-induced assembly is a simple and effective process for assembling highly aligned nanowires on stretchable substrates at the interface. Here, we demonstrate various methods such as striking method [73], contact printing method [78], and pressure release self-assembly [79] as an example. The contact printing process described herein is carried out in a specified direction on top of a receiver substrate on which a patterned etchant layer is printed using lithography, comprising a dense nanowire “lawn” [80, 81]. In this process, the nanowires are combed during the sliding process and leave the donor's substrate due to the van der Waals interaction with the surface of the receiver substrate, eventually leading to the direct transfer of the ordered nanowires to the receiver chip. Javey et al. developed contact printing methods to transfer conventional arrays of semiconductor nanowires directly from donor films to patterned receiver substrates. Figure 1.6a shows how this method can be used for assembly of single and high-density parallel nanowire arrays [78] on a substrate in a large scale. Surface chemical modification of the receiver substrate and weak interactions between the chemically unmodified NWs play an important role in the assembly process. The assembly of nanowire by knocking down method includes directional “knock-down” of a growth substrate, comprising a preprogrammed ordered dense “lawn” of nanowire arrays, for the measured in-place planarization of the vertical nanowire elements [82]. Figure 1.6b presents the schematic illustration of the percussion process reported by Patolsky and his collaborators, an initial work in this area. First, a vertical array of silicon nanowires with 4 μm length and 80 nm in diameter was obtained by top-down etching a silicon-insulator substrate. Next, a roller made of polydimethylsiloxane (PDMS), polytetrafluoroethylene, or another elastomer with different hardness and surface characteristics is manually rolled on the nanoarray substrate. In contrast to the contact printing process, the nanowire elements are grown directly on the final device substrate that is useful for the transfer from a donor substrate [82]. Lei and coworkers have also developed a two-step knockdown method that results in a high-density, highly ordered array of ZnO nanowires [83] in the *c*-axis direction on a flexible substrate. First, a controlled chemical vapor deposition (CVD) process is used initially to fabricate a vertical ZnO nanowire array perpendicular to the donor substrate surface. Subsequently, a manual contact printing process is used to knock down the nanowires and transfer them to the recipient substrate while maintaining the desired pattern [83]. As an efficient method of fabricating highly ordered nanowires, the strain release assembly method

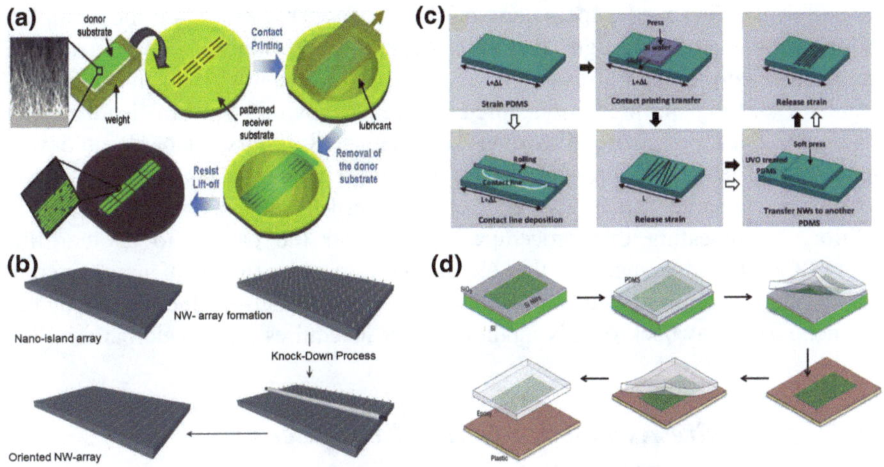

Fig. 1.6 A diagram of the NW assembly process. **a** NW contact printing. **b** Knockdown assembly method. **c** Strain relief NW assembly method. **a–c** are referenced by Refs. [73, 78, 79] Copyright 2008, 2010 and 2011 American Chemical Society. **d** Illustration of the transfer printing NW assembly method. Reproduced with permission from Ref. [75] Copyright 2007 Nature Publishing Group

involves first transferring the nanowires onto a compactable stretchable substrate, and then releasing the compacted substrate [79], resulting in the nanowires alignment in the lateral direction and the increase in the coverage area of the nanowires on the substrate. This method is applicable to any nanowires, synthesized by different methods or deposited under different conditions, and can be repeated several times to increase the ordering and density of the nanowires. The elasticity of the substrate and the static friction between the nanowires and the substrate are considered key factors in this assembly process. Zhu and his collaborators used silver nanowires and Si nanowires as template materials to demonstrate the versatility of this assembly approach. Figure 1.6c illustrates the assembly of nanowires in this process [79]. First, silver or silicon nanowires are transferred onto a compressed PDMS substrate by either contact line deposition or contact printing. After the removal of the growth substrate, the PDMS is coated with a layer of substantially ordered silver or Si nanowire film. When the PDMS film is released, the order and density of the nanowires also increase. In addition, further nanowire transfer can continue to increase the degree of order and density. As a novel spinning method, hand-spinning filament technique is used to prepare unidirectional ordered nanofibers [74]. The simplicity of the process is the most significant advantage. In addition, the diameter and surface morphology of fibers obtained by hand spinning are largely dependent on the PAD and solvent systems. Furthermore, Cao and his coworkers have developed a large-scale self-assembly method for varying the suspended single-walled carbon nanotubes (SWNTs) via conventional TiO_2 colloidal islands [84]. First, random SWNTs are layered on top of a small portion of the uniformly dispersed supernatant of TiO_2 nanoparticles. After the complete drying chapping of TiO_2 gel into microscale islands, random SWNTs can be straightened into aligned arrays. The transfer printing nanowire assembly method comprises shifting the premanufactured ordered nanowire structural features of a mold into a flat polymeric film over large areas [85]. First, a hard mold with topographical features is pressed against a polymeric material that covers the surface of the substrate, which creates a clear pattern in the polymeric material. In addition, by repeating this procedure can transfer the pattern to another film (Fig. 1.6d). The advantages of this technology are independent of nanomaterials and can be used for a wide range of polymeric materials including conjugated polymers. The drawback of this method is that it requires a premade pattern [75].

1.2.3 Nanowire Assemblies Induced by External Nanostructures

Properly applied materials can promote the ordering of some nanowires on the interface. The interaction between the nanowires and the supporting material promotes the alignment of the nanowires. Wu and Li studied co-assembly behavior such as disordered $Na_{0.44}MnO_2$ by using graphene oxide (GO) nanosheets (Fig. 1.7a–c) [86]. In the aqueous suspension of nanowires, the GO nanosheets are added, resulting in absorption of GO nanosheets onto to the surface of the nanowires by the interaction with the nanowires through hydrogen bonds, ionic dipoles, etc., which changed the surface properties of the nanowires and led to their development at the solution surface [86]. The adsorption of GO further stabilizes the colloidal solution by increasing the negative surface charge density of the nanowires. Moreover, the alignment of nanowires occurs as the concentration of nanowire reaches a critical value, proposed by Onsager's theory [87]. Yang et al. reported a method for self-assembly of gold nanowires with the diameter of 1.6 nm, and ultrahigh aspect ratios ($L/d > 500$) along the axes of multiwalled carbon nanotubes [88] are highlighted in Fig. 1.7. Strong hydrophobic and van der Waals interactions lead to self-assembly of allylamine-capped Au nanowires along the sidewalls of carbon nanotubes (CNTs) [89]. Adsorption and lattice orientation can guide the nanowires to form ordered films (Fig. 1.7e, f). Fibrous collagen was self-assembled into a highly ordered array on the mica surface by adsorption, surface diffusion, and nucleation, and grown into a two-dimensional lattice. In conclusion, interfaces offer an economical and efficient platform for nanowire assembly, but these are consistent with the past and current approaches to carbon nanotubes.

1.2.4 Assembly of 1D Nanostructures by External Fields

In order to modulate or control the structure, various physical forces like magnetic or electric fields are often applied. An extra field results in the self-assembly of

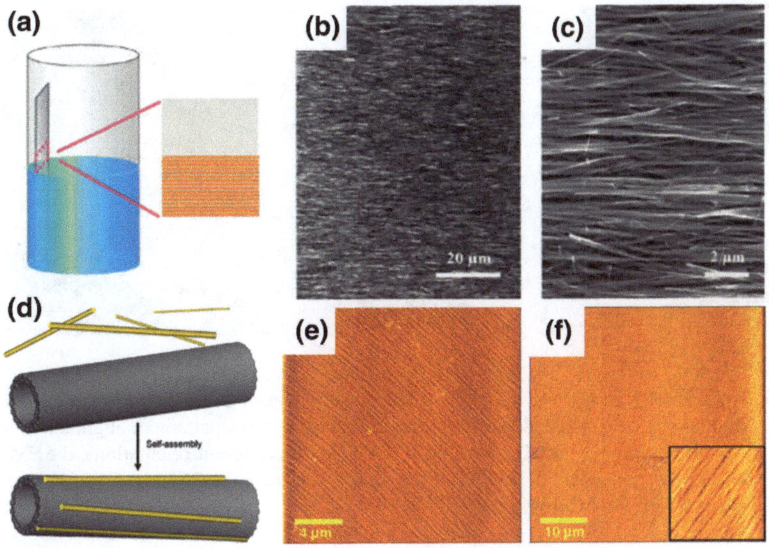

Fig. 1.7 Nanowire alignment with GO on a Si substrate. **a** Schematic showing the experimental setup: a piece of clean Si wafer was immersed in a well-dispersed nanowire/GO solution. After the solution is evaporated, the oriented nanowire pattern is deposited on a substrate having an orientation parallel to the horizontal liquid–substrate contact line. **b**, **c** Typical SEM image of the nanowire alignment pattern. Reproduced with permission from Ref. [76] Copyright 2009 American Chemical Society. **d** General scheme of self-assembly of Au nanowires on CNTs. Reproduced with permission from Ref. [80] Copyright 2011 American Chemical Society. **e**, **f** AFM images of the highly aligned collagen fibers on muscovite mica. [KCl] = 200 mM. The collagen concentration and incubation time of the mica were **e** 10 [mu] g/mL collagen and 1-min incubation and **f** 50 [mu] g/mL and 108-min incubation. *Insertion* 2 [mu] m scan. Reproduced with permission from Ref. [82] Copyright 2011 American Chemical Society

charged or magnetic material by influencing the growth and aggregation of nanobuilding blocks. Hence, external field is considered important for tuning the interaction between the nanomaterials, and consequently, efforts to control the transport and clustering properties have attracted a significant interest.

1.2.4.1 Magnetic Field-Assisted Assembly of 1D Nanostructures

A magnetic field is a force field that is obtained by moving electric charges, in response to a constant electric field or an intrinsic magnetic field related to the particle's spin. The magnetic field has appeared to be an interesting phenomenon for navigating migratory birds, homing animals, and even bacteria. Iron-containing cells are thought to provide a map of the earth's magnetic field lines for these organisms. These ferromagnetic crystals, like bar magnets, are polarized under geomagnetic guidance, which suggests that the magnetic field can influence the growth and directional aggregation of nanocrystals and lead to the self-assembly of

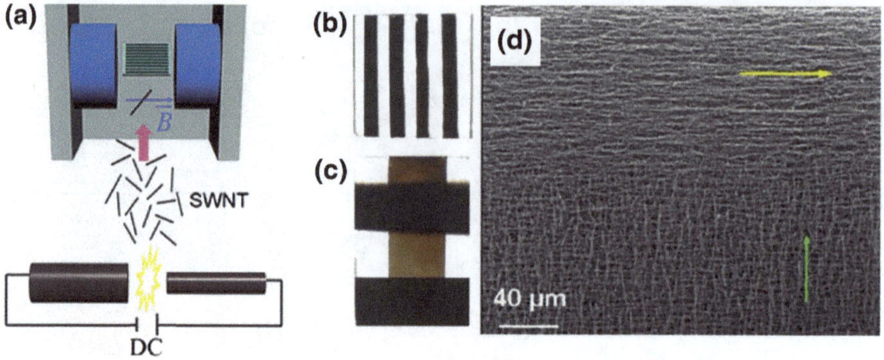

Fig. 1.8 **a** Schemes for making well-aligned and closely packed SWNT films on any substrate by means of a magnetic field-assisted arc discharge. **b** Photographs of patterned, aligned SWNT films on glass. **c** A photograph of the 3D structure of SWNTs with different orientations. **d** SEM images of aligned two-layer-deposited SWNT films with different orientations. Reproduced with permission from Ref. [92] Copyright 2010 WILEY-VCH

nanostructured materials [84, 88–91]. Chen and coworkers have demonstrated a method for fabricating highly ordered and closely aligned single-walled carbon nanotube (SWNT) films that combine arching growth with magnetic field guidance on a variety of substrate surfaces, including flexible plastics [91]. This method includes various phenomena such as the diffusion, arc discharge growth, and in situ formation of SWNTs grown under a magnetic field, and finally, these neat SWNTs are deposited on the substrate in the desired orientation and direction. The assembly process of well-aligned and closely packed SWNT film on a substrate is presented in Fig. 1.8a. By this technique, orientation, location, area, thickness, and hierarchical structure can be controlled. Moreover, by controlling the deposition parameters, evaporation coating technology can make a neat SWNT film to obtain the required geometry. Thus, by using a masking technique in a vacuum evaporation coating, structures of various shapes consisting of ordered SWNTs can be prepared directly in a given region [91]. As the orientation of the ordered SWNT thin film is determined solely by the associated magnetic field–substrate interaction, the sorting orientation of SWNTs can be easily altered by rotating the substrate. Thus, a multilayer three-dimensional structure (Fig. 1.8b–d) with various orientations for each layer can be easily prepared by adjusting the direction and position of the film/mask [92].

Moreover, Despina et al. developed a simple method for magnetic-induced formation, positioning and assembling of magnetic nanowires in a polymer film. By changing the duration of applied magnetic field, the positions and dimensions of particular layers in the polymer matrix could be controlled in combination with the evaporation dynamics [93]. The formation scheme shows the production of the well-defined film by placing nanobuilding blocks with certain concentration to an external magnetic field (Fig. 1.9b–e). In addition to the applied magnetic field, the polymer matrix also plays a key role in the process of developing nanowires at a

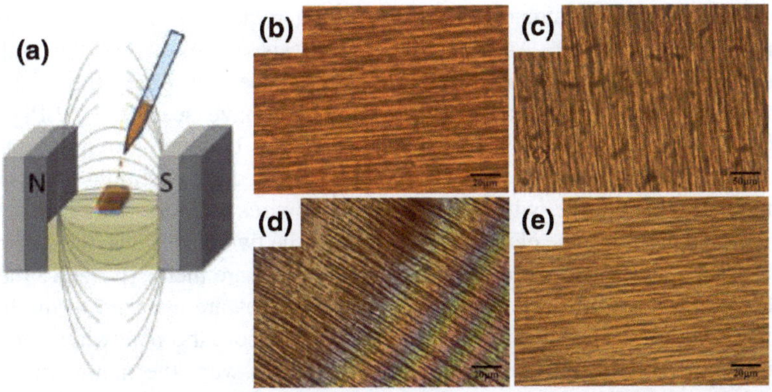

Fig. 1.9 **a** Flowchart of the manufacturing method. **b–e** Optical microscopic images of various colloidal NP/polymer modules (NW made of NP). **b** 1 wt% 99 wt% γ-Fe_2O_3 of poly(methyl methacrylate) (PMMA). **c** 1 wt% 99 wt% γ-Fe_2O_3 of polystyrene (PS). **d** 2 wt% 98 wt% of γ-Fe_2O_3 of poly[3-3′ (vinylcarbazole)] (PVK). **e** 1 wt% Fe_2O_3–TiO_2 heterostructure, synthesized in our laboratory, 99 wt% of PEMMA. Reproduced with permission from Ref. [94] Copyright 2010 American Chemical Society

particular depth of the film, since it initiates the aggregation process in the initial stage of growth and provides a viscosity gradient [94] inside the film. In particular, an ordered array of one-dimensional magnetic nanowires having a magnetic anisotropic behavior based on the nanoparticles in a plastic film can be prepared to be sufficiently thick that it can exist independently.

1.2.4.2 Electric Field-Assisted Assembly of 1D Nanostructures

In the presence of electric field, the nanowires can be positioned by introducing them in dielectric medium that strongly affects the charged object [95–101]. In dielectrophoresis (DEP) phenomenon, when the dielectric element is subjected to a nonuniform electric field (often AC electric field) they experience force exerted by electric field (normally AC electric field), providing a method to attract nanowires onto predesigned electrodes. Dielectrophoresis can be compared with the phenomenon of electrophoresis, in which suspended building block motion is created by the action of an electrostatic field on the charged building blocks [102]. Due to the separation of the surface charge of the nanowires, an alternating electric field can cause polarization of the nanowires. When the nanowires are more polarized than the dielectrics, the nanowires are subjected to the dielectrophoretic force and undergo a net shift in the direction of the increasing field strength that usually occurs at the edges of the electrodes. Thus, an electric field-assisted assembly method is used to arrange the nanowires, typically using a dielectrophoretic force to direct the nanowires toward the high field areas [101, 103]. DEP can be tuned to accommodate a wide variety of conductive nanowires of various materials and can accommodate any substrate, including those that require cryogenic processes, as well as flexible substrate. As the dielectrophoresis assembly has the ability to accurately position the nanowire nanotubes on the substrate, the method is becoming increasingly popular [104–107]. As we have seen, the DEP process depends on the nanowire density, dielectric medium, bias field strength, and bias duration [108]. First, the spacing between the two electrodes should be similar to the length of the nanowire, which allows the nanowires to shorten the spacing after alignment. On applying the electric field between the two electrodes, the electrically neutral regions of the nanowire become polarized that are then exposed to the DEP forces. Secondly, in order to ensure that the nanowires are attached to the highest gradient of the electric field at the edge of the electrode, the nanowires should be more conductive than the dielectric medium. In this way, the nanowires can be accurately assembled and positioned in complex devices. Freer et al. developed a promising approach to nanowire assembly that combines the accuracy and large-scale coverage of boron-doped silicon nanowire self-assembly with an unprecedented combination [106]. The optical dark-field and deep ultraviolet (DUV) images of nanowires after the complete assembly onto electrodes on a wafer-scale quartz are presented in Fig. 1.10a–c. Nanowires with a pitch less than the characteristic decay length of the AC field originating from the electrodes are polarized after the nanowire suspension flows through the channels. Under the dielectrophoretic force, these polarized nanowires are attracted to the electrodes, the magnitude of which depends on the frequency, voltage, electrode geometry, and material properties. Furthermore, the assembly process of the nanowires depends on the balance of hydrodynamics, dielectrophoretic, and the classical bilayer interaction between the nanowires and the surface. Thus, individual nanowires can have a high probability of assembly on the electrodes by careful control of the dielectrophoretic force and hydrodynamics [108].

It can be expected that the area where individual nanowire assemblies are present is limited only by the ability to maintain the uniformity of hydrodynamics and electrophoretic forces. Nanotubes can also be assembled in devices like nanowires by DEP processes [109–111]. For example, Khondaker and coworkers published on a single-walled carbon nanotube (SWNT) assembly process driven by DEP at a probe station [112]. First, a small drop of SWNT solution is dropped onto a chip containing an array of electrodes, with an AC voltage of 300 kHz and 5 V_{p-p} between the source and the drain using a signal generator for 30 s. The nanotubes show the translator motion along the electric field gradient and also show alignment in the direction of the electric field lines due to the interaction of induced dipole moment of the nanotubes with the strong electric field [88]. Figure 1.10d–i presents a typical DEP assembly by varying the concentration of the SWNT solution with a simple dilution of the original solution using DI (deionized) water. Different SWNT solution concentrations (e.g., 0.08, 0.34, 1.67, and 3.4 [mu] g/mL) are used to increase the SWNT density (from 1 to 30/[mu] m, respectively, corresponding to Fig. 1.10d–g). At low density, almost all of the nanotubes are well aligned and parallel to each other. Conversely, about 90% of the nanotubes are aligned in a range of ±10° along the longitudinal axis at high concentrations. As it is clear from the figure, the density of the SWNTs in the channel can be varied from 1 to 30 SWNT/m by simply changing the concentration of the SWNT solution [88].

1.2.5 Nanowire Assemblies by Microfluidic Flow

From the nineteenth century, river water was used to transport logs to the downstream sawmill. This ancient method of transporting logs has led to a new approach for the assembly of nanowires. Even at low loading, one-dimensional nanowires can be greatly affected by the rheological properties of the suspension due to its large aspect ratio [113]. The microflows allow the nanowires alignment by linking a fluid arrangement with a surface patterning technique [109, 114–119]. Firstly, a homogeneous and stable suspension of nanowires is prepared, which is allowed to pass through the microfluid channels, resulting in the aggregation of nanowires as a whole. The nanowires are arranged in the direction of flow driven by shear forces, and the assembled nanowire density receives control from the concentration of nanowires in the suspension and the flow time. By the flow rate, the alignment uniformity can be controlled easily, since high flow rates produce a greater shearing force and thus make the nanowires more align. Similarly, by the flow duration, the average nanowire surface coverage can be controlled. Furthermore, the average density of NWs can be calculated by dividing the average number of NWs at any cross section of the channel by the width of the channel. With the help of layer-by-layer process, complex geometries like crossed nanowire arrays can be formed. Lieber group reported a method for the hierarchical assembly of one-dimensional nanostructures into well-defined functional networks using microfluidic flow [120]. Figure 1.11a shows a nanowire assembly array with microflow assistance. First, the nanowires are suspended in an ethanol solution. Allowing the suspension to flow through the channel structure formed by the polydimethylsiloxane template, the one-dimensional nanowires are aligned in the microfluidic, and the nanowires are layered assembled on the flattened substrate by separation and spatial position control (Fig. 1.11b). Crossed arrays of nanowires can also be prepared by the layer-by-layer assembly, using different flow directions for successive steps (Fig. 1.11c). Thus, parallel and crossed arrays of nanowires can be obtained by the assembly process of the individual flow (Fig. 1.11b) and the continuous cross-flow (Fig. 1.11c) is described below. Kim and his collaborators reported on an assembly method [121] driven by microfluidics through rheological control to achieve macroscopic nanowire alignment [122]. Arrangement involves primarily a severe shearing process performed by a rheometer. In general, the shear rate is >100 s^{-1} and the shear time is greater than 1 s. It is found that higher shear rates usually results in better alignment. In addition, after the steady state is reached, the longer shear time no longer has an effect on alignment. Ordered nanostructures are formed by this intense shearing action, which can be controlled reversibly by different shear rates.

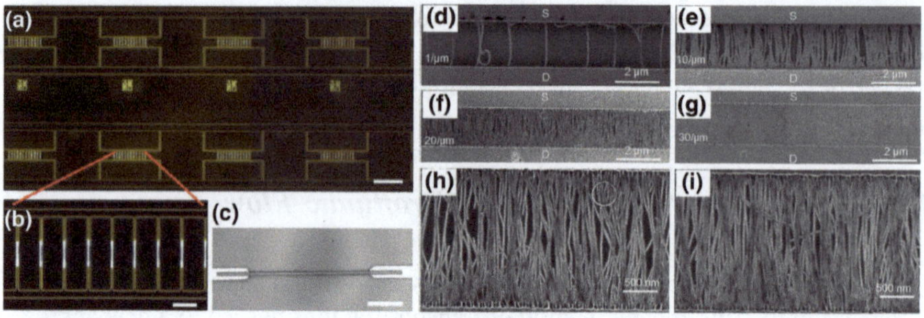

Fig. 1.10 The nanowires are manipulated by the field region or the typical DEP assembly. **a–c** undergoes a complete process of assembling the nanotube optical dark-field image and the DUV image on a quartz substrate. **d** is about 1 SWNT/m; **e** 10 SWNT/m; **f** 20 SWNT/[mu]; **g** 30 SWNT/μm. "S" and "D" in **d–i** denote source and drain. Reproduced with permission from Ref. [95] Copyright 2011 American Chemical Society

1.2.6 Nanowire Assemblies by Bubble Blowing Process

Blown film extrusion is the most common method that involves extruding a molten polymer and expanding it to obtain a balloon, especially used for the mass production of the plastic film, and can be collapsed and slit to form continuous flat films. The blowing bubble assembly method is a versatile and extensible assembly method. Recently, this method has been developed for assembly of nanowires and nanotubes [123, 124] with a single ordered and controlled density by controlling the extension of the bubbles formed by a homogeneous polymer suspension containing one-dimensional nanomaterials. Figure 1.12 illustrates the main steps. Firstly, the functional nanowires are diffused in a precise concentration of the polymer and form a homogeneous, stable suspension (Fig. 1.12a). Subsequently, the polymer suspension is expanded into a bubble using a circular die at a controlled pressure P and dilation rate. Here, a stable vertical expansion can be achieved by using an applied vertical force F (Fig. 1.12b). The transfer of the bubble film onto the substrates or the open structure is presented in Fig. 1.12c [125]. An outstanding alignment of the Si nanowires is perceived in the bubble film (BBF) obtained using epoxy suspension with different concentrations of Si nanowires (0.01–0.22 wt%) (Fig. 1.12d–g). Qualitatively, the shear stress related with the suspension fleeting through the peripheral edge of the die allows the high-aspect-ratio nanowires to be aligned along the principal direction of the strain in the polymer flow. When the bubbles expand in the vertical direction, the nanowire orientation in the BBF should always be in the upward (warp) direction as the nanowire suspension is replenished at the top of the die, which is consistent with the observed optical image. It is important to expand in a definite direction for the nanowires that obtain the same orientation in a large scale and for the overall orientation to be fixed in absolute observation during the transfer to the film [125].

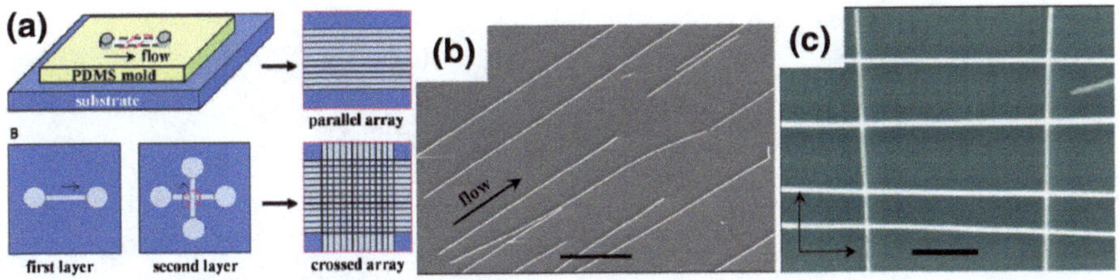

Fig. 1.11 Schematic representation of the fluid passage structure in fluid assembly. **a** A channel formed when the PDMS mold is in contact with a flat substrate. Let the NW suspension flow through the channel at a constant flow rate and duration, and the NW will be assembled. **b** A parallel array of NWs as observed after the removal of the PDMS mold in the film flow direction. **c** A multiple-interleaved NW array in which the flow direction can be sequentially changed in the layer-by-layer assembly process. Reproduced with permission from Ref. [34] Copyright 2001 The American Association for the Advancement of Science

1.2.7 Nanowire Assemblies by Electrospinning

Electrospinning is a highly versatile method [126–129] that utilizes strong electric field forces to process a solution or a polymer-based melt into continuous fibers with diameters between a few nanometers and a few micrometers [130–135]. Electrospinning mainly utilizes electrostatic repulsion between surface charges to shrink the diameter of viscoelastic jet or glass filaments. Under the influence of the strong electrostatic field, the electrospinning fibers are assembled in the axial direction. The composite material can be generated by adding additional space constraints to the polymer chain. Figure 1.14a presents a typical arrangement of electrospinning experiment in which the polymer solution or melt is pumped through a thin nozzle with an inner diameter of the order of 100 μm. The nozzle also acts as an electrode with a strong electric field of 100–500 kV m^{-1}, whereas the distance between opposing electrodes is about 10–25 cm. Yu group has developed a method in which an independent flexible surface-enhanced Raman scattering (SERS) substrate has been synthesized by electrospinning using polyvinyl alcohol (PVA) and dimer or ordered aggregates of Au, which are assembled on chain-like nanofibers (Fig. 1.13b) [135]. Figure 1.13c shows the formation of Ag/PVA nanofibers by electrospinning, indicating a three-dimensional lattice structure containing a large number of randomly deposited fibers, where the individual fibers have a high aspect ratio and a smooth surface. Figure 1.13d–g show that the Ag aggregates are fixed in the PVA matrix and are assembled along the axial direction into some ordered linear chain structure. The Ag/PVA nanofibers have a diameter of about 170 nm and a length of several millimeters. The Au nanoparticle aggregates results in a chain-like array in the PVA matrix after the electrospinning process, as compared to the solution in which some aggregates were randomly distributed before electrospinning [135]. Recently, Yu et al. reported a method for large-scale preparation of independent, stable Au NR/PVA nanofiber mats by electrospinning of Au nanorods (NRs) assembly [136]. The Au

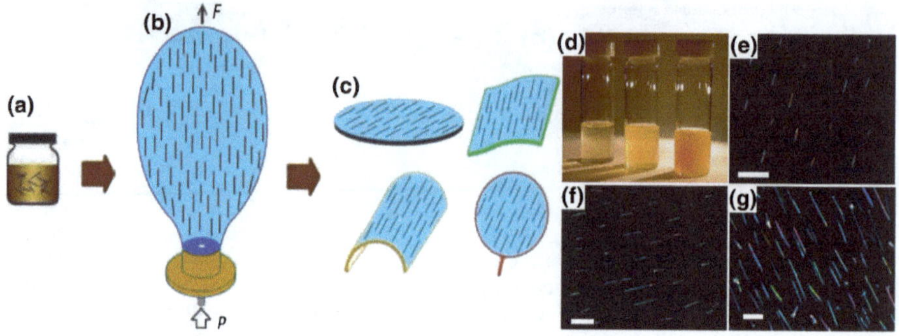

Fig. 1.12 Blowing the bubble film (BBF) process diagram. **a** NW/NT-polymer suspension, **b** expansion of the bubbles on a circular die, and **c** transfer to films on wafers, plastics, curved planes, and open structures. Nitrogen gas with a pressure of *P* is blown through the die and the bubble is gradually increased from the NW/NT epoxy suspension (*dark blue*) at the top of the die, while a stable vertical force *F* is controlled by connecting the controllable speed motor The wire loop is applied to it. The *black line* represents the aligned NWs/NTs embedded in the bubble film. **d–g** Control of the arrayed NW density in the BBFs. **d** 0.01, 0.03, and 0.15 wt% (from *left* to *right*) of an epoxy resin suspension containing Si NWs. **e–g** of dark-field optical images of 0.01% **e**, 0.03 **f**, and 0.15 **g** wt% Si NWs-BBFs, respectively. The scale lengths in **e**, **f**, and **g** are 50, 20, and 10 mm, respectively. Reproduced with permission from Ref. [125] Copyright 2007 Nature Publishing Group

NRs are assembled in the PVA fibers and along the axial direction of the fibers. The spacing of adjacent nanorods, containing "head-to-head" and "side-to-side" distance, can be changed by varying the concentration of Au NRs in the PVA solution. In addition, Tracy and his coworkers fabricated long-range ordered nanorods on a macroscopic scale by using poly(ethylene oxide) (PEO) fiber membranes obtained by electrospinning [137]. As a result, nanowires can be assembled in a well-orientated manner, which is expected to be excellent in the future. In order to obtain electrospun fiber bundles arranged neatly or in a specific pattern, electrodes or rotating collectors which are parallel to each other (at a certain pitch in a certain angle) have been used. The ordered fibers can be prepared by mixing the magnetic nanoparticles with the macromolecular polymer [133, 138, 139], through applying an external magnetic field. In the process of heterogeneous electrospinning, the design of the collector electrode is very important for obtaining the oriented nanofibers.

At the same time, the electrospinning process has been also evolved to arrange the nanofibers [140] using rotating electrodes [141, 142] or electrostatic forces from a particular counterelectrode. Xia's group reported an improved electrospinning process [143] for generating unidirectional nanofibers containing different components. In order to attain a parallel array of nanowires, the conventional collector is divided into two blocks and separated by a gap. When the Au electrodes on the insulator are mated and alternately grounded (Fig. 1.14a), a double-layer nanowire mesh (Fig. 1.14b) can be obtained. In addition, the nanowires can be aligned by modifying the collectors by a magnetic field or an electric field. More recently,

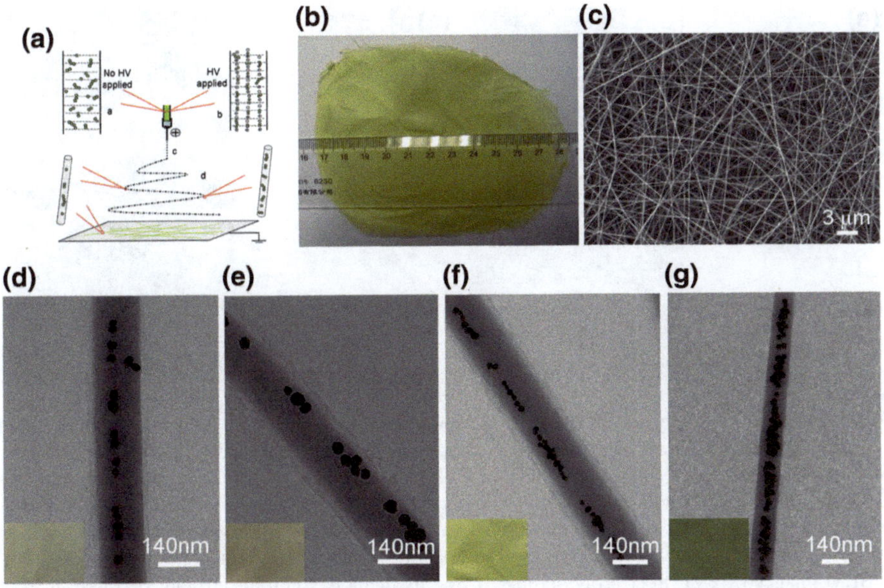

Fig. 1.13 **a** A laboratory electrospinning setup with vertical electrodes. **b** Photomicrographs of PVA nanofiber mats (PVA/Ag molar ratio 530:3) obtained by electrospinning for 1 h. **c** Typical SEM images of Ag/PVA nanofiber mats. **d–g** PVA/Ag molar ratio of **d** 530:1, **e** 530:2, **f** 530:3, and **g** 530:4, respectively. The inset is a photograph of the corresponding Ag/PVA nanofiber mat. Reproduced with permission from Ref. [129] Copyright 2009 American Chemical Society

Yang et al. has reported a magnetic field-assisted electrospinning process [133] for preparing ordered nanowires. Figure 1.14c is a schematic diagram of an electrospinning device incorporating a magnetic field-modified collector. Unidirectional aligned PLGA fibers after 120 min of the collection are presented in Fig. 1.14d. The magnetic field causes an additional force of the spray and increases the velocity of the spray reaching the film. The advantage of using an external magnetic field in place of a parallel auxiliary electrode to produce ordered nanofibers is that the resulting array is held on a thick fiber membrane. The researchers have also used high-speed rollers as the collector in the heterogeneous electrospinning in order to replace the parallel electrodes. By using a rotating disk collector with a sharp edge (diameter 200 mm), ordered nanofiber scaffolds can be prepared by electrospinning process (Fig. 1.14e, f) [130]. Figure 1.14f is an optical image of an electrospun nanofiber which shows that the three-dimensional fiber mesh consists of fibers having a diameter in the range of 200–800 nm and the thickness of the nanowire film is about 0.5 mm. Most of the fibers are aligned along the longitudinal axis, which form a unique array of topographies. Gu et al. have described a heterogeneous electrospinning process [130] which combines a rotating collector and a parallel electrode of the two collector-modifying methods. Figure 1.14g, h shows an improved electrospinning device. In most cases, nanotubes are very similar to nanowires, and the assembly methods of nanowires are consistent with those

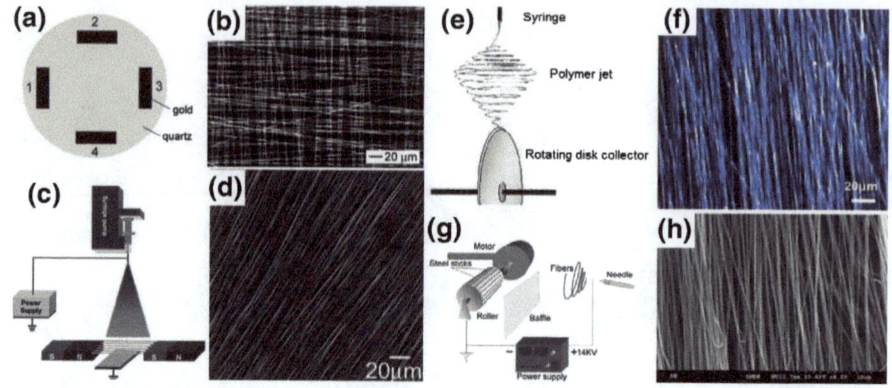

Fig. 1.14 **a**, **b** Assembly of an electrospun nanowire equipped with a gold electrode-modified collector. **a** Schematic view of a test pattern comprising four gold electrodes on a quartz substrate. **b** An optical microscopic image of a mesh of PVP nanowires collected in the central region of the electrode. Reproduced with permission from Ref. [127] Copyright 2004 WILEY-VCH **c**, **d** is provided with an electrospinning nanowire assembly process using a magnetic field-modified collector. **c** A schematic representation of an apparatus for preparing an ordered nanofiber film. **d** SEM images of PVP fiber arrays prepared by this method. Reproduced with permission from Ref. [133] Copyright 2010 WILEY-VCH **e**, **f** with the high-speed rotor-modified collector of electrospinning nanowires assembly process. **e** A schematic representation of an apparatus for preparing ordered nanofiber scaffolds using an electrospinning process equipped with a rotating disk collector. **f** Optical microscopic images of ordered organic nanofiber scaffolds. Reproduced with permission from Ref. [130] Copyright 2004 Elsevier. **g**, **h** An electrospinning nanowire assembly process with a high-speed insulator-coated collector coated with a parallel iron bar. **g** A modified apparatus for performing an electrospinning operation. **h** SEM images of fibers collected with runners and baffles covered with parallel iron bars. Reproduced with permission from Ref. [138] Copyright 2007 WILEY-VCH

methods that have been or are being used on nanotubes. In fact, the boundaries of different nanowire assembly methods are very fuzzy. Many assembly strategies combine more than two kinds of specific methods, such as LB technology is the combination of mechanical force and interface effect. Most of the time, we sort the strategies by the method. In some cases, we define the strategy as mechanical force method because contact printing, striking, and pressure release methods are mainly force-related assembly methods.

References

1. Law M, Goldberger J, Yang P (2004) Semiconductor nanowires and nanotubes. Annu Rev Mater Res 34:83–122
2. Pan H, Feng YP (2008) Semiconductor nanowires and nanotubes: effects of size and surface-to-volume ratio. ACS Nano 2(11):2410–2414
3. Xia Y, Yang P, Sun Y, Wu Y, Mayers B, Gates B, Yin Y, Kim F, Yan H (2003) One-dimensional nanostructures: synthesis, characterization, and applications. Adv Mater 15 (5):353–389

4. Hochbaum AI, Yang P (2009) Semiconductor nanowires for energy conversion. Chem Rev 110(1):527–546
5. Yang P, Yan R, Fardy M (2010) Semiconductor nanowire: what's next? Nano Lett 10 (5):1529–1536
6. Tang J, Huo Z, Brittman S, Gao H, Yang P (2011) Solution-processed core-shell nanowires for efficient photovoltaic cells. Nat Nanotech 6(9):568–572
7. Qin Y, Wang X, Wang ZL (2008) Microfibre-nanowire hybrid structure for energy scavenging. Nature 451(7180):809–813
8. Pauzauskie PJ, Radenovic A, Trepagnier E, Shroff H, Yang P, Liphardt J (2006) Optical trapping and integration of semiconductor nanowire assemblies in water. Nat Mater 5(2): 97–101
9. Xu S, Qin Y, Xu C, Wei Y, Yang R, Wang ZL (2010) Self-powered nanowire devices. Nat Nanotech 5(5):366–373
10. Feng M, Zhang M, Song J-M, Li X-G, Yu S-H (2011) Ultralong silver trimolybdate nanowires: synthesis, phase transformation, stability, and their photocatalytic, optical, and electrical properties. ACS Nano 5(8):6726–6735
11. Cui X, Yu SH, Li L, Biao L, Li H, Mo M, Liu XM (2004) Selective synthesis and characterization of single-crystal silver molybdate/tungstate nanowires by a hydrothermal process. Chem A Eur J 10(1):218–223
12. Qian H-S, Yu S-H, Luo L-B, Gong J-Y, Fei L-F, Liu X-M (2006) Synthesis of uniform Te@ carbon-rich composite nanocables with photoluminescence properties and carbonaceous nanofibers by the hydrothermal carbonization of glucose. Chem Mater 18(8):2102–2108
13. Qian H-S, Yu S-H, Gong J-Y, Luo L-B, Fei L-F (2006) High-quality luminescent tellurium nanowires of several nanometers in diameter and high aspect ratio synthesized by a poly (vinyl pyrrolidone)-assisted hydrothermal process. Langmuir 22(8):3830–3835
14. Wang K, Liang H-W, Yao W-T, Yu S-H (2011) Templating synthesis of uniform Bi 2 Te 3 nanowires with high aspect ratio in triethylene glycol (TEG) and their thermoelectric performance. J Mater Chem 21(38):15057–15062
15. Liu J-W, Chen F, Zhang M, Qi H, Zhang C-L, Yu S-H (2010) Rapid microwave-assisted synthesis of uniform ultralong Te nanowires, optical property, and chemical stability. Langmuir 26(13):11372 11377
16. Pauzauskie PJ, Yang P (2006) Nanowire photonics. Mater Today 9(10):36–45
17. Trentler TJ, Hickman KM, Goel SC, Viano AM (1995) Solution-liquid-solid growth of crystalline III-V semiconductors: an analogy to vapor-liquid-solid growth. Science 270 (5243):1791
18. Dai H, Wong EW, Lu YZ, Fan S, Lieber CM (1995) Synthesis and characterization of carbide nanorods. Nature 375(6534):769–772
19. Yang P, Lieber CM (1996) Nanorod-superconductor composites: a pathway to materials with high critical current densities. Science 273(5283):1836
20. Goldberger J, Hochbaum AI, Fan R, Yang P (2006) Silicon vertically integrated nanowire field effect transistors. Nano Lett 6(5):973–977
21. Law M, Greene LE, Johnson JC, Saykally R, Yang P (2005) Nanowire dye-sensitized solar cells. Nat Mater 4(6):455–459
22. Yuan J, Liu X, Akbulut O, Hu J, Suib SL, Kong J, Stellacci F (2008) Superwetting nanowire membranes for selective absorption. Nat Nanotechnol 3(6):332–336
23. Tang SK, Derda R, Mazzeo AD, Whitesides GM (2011) Reconfigurable self-assembly of mesoscale optical components at a liquid-liquid interface. Adv Mater 23(21):2413–2418
24. Grzybowski BA, Wilmer CE, Kim J, Browne KP, Bishop KJ (2009) Self-assembly: from crystals to cells. Soft Matter 5(6):1110–1128
25. Triplett DA, Quimby LM, Smith BD, Hernández Rodríguez D, St. Angelo SK, González P, Keating CD, Fichthorn KA (2009) Assembly of gold nanowires by sedimentation from suspension: experiments and simulation. J Phys Chem C 114(16):7346–7355
26. Sakakibara K, Hill JP, Ariga K (2011) Thin-film-based nanoarchitectures for soft matter: controlled assemblies into two-dimensional worlds. Small 7(10):1288–1308

27. Hu J, Odom TW, Lieber CM (1999) Chemistry and physics in one dimension: synthesis and properties of nanowires and nanotubes. Acc Chem Res 32(5):435–445
28. Acharya S, Hill JP, Ariga K (2009) Soft Langmuir–Blodgett technique for hard nanomaterials. Adv Mater 21(29):2959–2981
29. Tao AR, Huang J, Yang P (2008) Langmuir–Blodgettry of nanocrystals and nanowires. Acc Chem Res 41(12):1662–1673
30. Jiang L, Dong H, Hu W (2011) Controlled growth and assembly of one-dimensional ordered nanostructures of organic functional materials. Soft Matter 7(5):1615–1630
31. Yan H, Choe HS, Nam S, Hu Y, Das S, Klemic JF, Ellenbogen JC, Lieber CM (2011) Programmable nanowire circuits for nanoprocessors. Nature 470(7333):240–244
32. Lu W, Lieber CM (2007) Nanoelectronics from the bottom up. Nat Mater 6(11):841–850
33. Huang Y, Duan X, Cui Y, Lauhon LJ, Kim K-H, Lieber CM (2001) Logic gates and computation from assembled nanowire building blocks. Science 294(5545):1313–1317
34. Huang Y, Duan X, Wei Q, Lieber CM (2001) Directed assembly of one-dimensional nanostructures into functional networks. Science 291(5504):630–633
35. Kumar P (2010) Directed self-assembly: expectations and achievements. Nanoscale Res Lett 5(9):1367–1376
36. Zhuang X, Ning C-Z, Pan A (2012) Composition and bandgap-graded semiconductor alloy nanowires. Adv Mater 24(1):13–33
37. Wang MC, Gates BD (2009) Directed assembly of nanowires. Mater Today 12(5):34–43
38. Gratton SE, Williams SS, Napier ME, Pohlhaus PD, Zhou Z, Wiles KB, Maynor BW, Shen C, Olafsen T, Samulski ET (2008) The pursuit of a scalable nanofabrication platform for use in material and life science applications. Acc Chem Res 41(12):1685–1695
39. Liu JW, Xu J, Liang HW, Wang K, Yu SH (2012) Macroscale ordered ultrathin telluride nanowire films, and tellurium/telluride hetero-nanowire films. Angew Chem Int Ed 51 (30):7420–7425
40. Liu J-W, Zhu J-H, Zhang C-L, Liang H-W, Yu S-H (2010) Mesostructured assemblies of ultrathin superlong tellurium nanowires and their photoconductivity. J Am Chem Soc 132 (26):8945–8952
41. Kim F, Kwan S, Akana J, Yang P (2001) Langmuir–Blodgett nanorod assembly. J Am Chem Soc 123(18):4360–4361
42. Tao A, Kim F, Hess C, Goldberger J, He R, Sun Y, Xia Y, Yang P (2003) Langmuir–Blodgett silver nanowire monolayers for molecular sensing using surface-enhanced Raman spectroscopy. Nano Lett 3(9):1229–1233
43. Wang D, Chang Y-L, Liu Z, Dai H (2005) Oxidation resistant germanium nanowires: bulk synthesis, long chain alkanethiol functionalization, and Langmuir–Blodgett assembly. J Am Chem Soc 127(33):11871–11875
44. Patla I, Acharya S, Zeiri L, Israelachvili J, Efrima S, Golan Y (2007) Synthesis, two-dimensional assembly, and surface pressure-induced coalescence of ultranarrow PbS nanowires. Nano Lett 7(6):1459–1462
45. Shevchenko EV, Talapin DV, Kotov NA, O'Brien S, Murray CB (2006) Structural diversity in binary nanoparticle superlattices. Nature 439(7072):55–59
46. Zheng R, Popov YO, Witten TA (2005) Deposit growth in the wetting of an angular region with uniform evaporation. Phys Rev E 72(4):046303
47. Gotkis Y, Ivanov I, Murisic N, Kondic L (2006) Dynamic structure formation at the fronts of volatile liquid drops. Phys Rev Lett 97(18):186101
48. Maheshwari S, Zhang L, Zhu Y, Chang H-C (2008) Coupling between precipitation and contact-line dynamics: multiring stains and stick-slip motion. Phys Rev Lett 100(4):044503
49. Zhang C, Zhang X, Zhang X, Fan X, Jie J, Chang JC, Lee CS, Zhang W, Lee ST (2008) Facile one-step growth and patterning of aligned squaraine nanowires via evaporation-induced self-assembly. Adv Mater 20(9):1716–1720
50. Ming T, Kou X, Chen H, Wang T, Tam HL, Cheah KW, Chen JY, Wang J (2008) Ordered gold nanostructure assemblies formed by droplet evaporation. Angew Chem 120(50): 9831–9836

51. Li H, Cheng C, Li X, Liu J, Guan C, Tay YY, Fan HJ (2012) Composition-graded Zn x Cd1−x Se@ ZnO core–shell nanowire array electrodes for photoelectrochemical hydrogen generation. J Phys Chem C 116(5):3802–3807
52. Wu JH, Xu TZ, Ang SG, Xu Q-H, Xu GQ (2011) Radially oriented anthracene nanowire arrays: preparation, growth mechanism, and optical fluorescence. Nanoscale 3(4):1855–1860
53. Kalsin AM, Fialkowski M, Paszewski M, Smoukov SK, Bishop KJ, Grzybowski BA (2006) Electrostatic self-assembly of binary nanoparticle crystals with a diamond-like lattice. Science 312(5772):420–424
54. Lin X-M, Cui Y, Xu Y-H, Ren B, Tian Z-Q (2009) Surface-enhanced Raman spectroscopy: substrate-related issues. Anal Bioanal Chem 394(7):1729–1745
55. Du Y-P, Zhang Y-W, Yan Z-G, Sun L-D, Yan C-H (2009) Highly luminescent self-organized sub-2-nm EuOF nanowires. J Am Chem Soc 131(45):16364–16365
56. Huo Z, Tsung C-K, Huang W, Zhang X, Yang P (2008) Sub-two nanometer single crystal Au nanowires. Nano Lett 8(7):2041–2044
57. Xu W, Leeladhar R, Tsai Y-T, Yang E-H, Choi C-H (2011) Evaporative self-assembly of nanowires on superhydrophobic surfaces of nanotip latching structures. Appl Phys Lett 98(7):073101
58. Kim TY, Kwon SW, Park SJ, Yoon DH, Suh KS, Yang WS (2011) Self-organized graphene patterns. Adv Mater 23(24):2734–2738
59. Kwon SW, Byun M, Yoon DH, Park J-H, Kim W-K, Lin Z, Yang WS (2011) Simple route to ridge optical waveguide fabricated via controlled evaporative self-assembly. J Mater Chem 21(14):5230–5233
60. Kwon SW, Yoon DH, Yang WS (2011) A simple route of ordered high quality mesoscale stripe polymer patterns. Soft Matter 7(5):1682–1685
61. Hong SW, Jeong W, Ko H, Kessler MR, Tsukruk VV, Lin Z (2008) Directed self-assembly of gradient concentric carbon nanotube rings. Adv Funct Mater 18(14):2114–2122
62. Duan H, Wang D, Kurth DG, Möhwald H (2004) Directing self-assembly of nanoparticles at water/oil interfaces. Angew Chem Int Ed 43(42):5639–5642
63. Wang Z, Bao R, Zhang X, Ou X, Lee CS, Chang JC, Zhang X (2011) One-step self-assembly, alignment, and patterning of organic semiconductor nanowires by controlled evaporation of confined microfluids. Angew Chem Int Ed 50(12):2811–2815
64. Reincke F, Hickey SG, Kegel WK, Vanmaekelbergh D (2004) Spontaneous assembly of a monolayer of charged gold nanocrystals at the water/oil interface. Angew Chem Int Ed 43 (4):458–462
65. Tsai Y-T, Xu W, Yang E-H, Choi C-H (2010) Self-assembly of nanowires at three-phase contact lines on superhydrophobic surfaces. Nanosci Nanotechnol Lett 2(2):150–156
66. Zhang C, Zhang X, Zhang X, Ou X, Zhang W, Jie J, Chang JC, Lee CS, Lee ST (2009) Facile one-step fabrication of ordered organic nanowire films. Adv Mater 21(41):4172–4175
67. Van Hameren R, Schön P, Van Buul AM, Hoogboom J, Lazarenko SV, Gerritsen JW, Engelkamp H, Christianen PC, Heus HA, Maan JC (2006) Macroscopic hierarchical surface patterning of porphyrin trimers via self-assembly and dewetting. Science 314(5804): 1433–1436
68. Knowles TP, Oppenheim TW, Buell AK, Chirgadze DY, Welland ME (2010) Nanostructured films from hierarchical self-assembly of amyloidogenic proteins. Nat Nanotechnol 5(3):204–207
69. Shi HY, Hu B, Yu XC, Zhao RL, Ren XF, Liu SL, Liu JW, Feng M, Xu AW, Yu SH (2010) Ordering of disordered nanowires: spontaneous formation of highly aligned, ultralong Ag nanowire films at oil–water–air interface. Adv Funct Mater 20(6):958–964
70. Deegan RD, Bakajin O, Dupont TF, Huber G, Nagel SR, Witten TA (1997) Capillary flow as the cause of ring stains from dried liquid drops. Nature 389(6653):827–829
71. Bernard C, Aime J-P, Marsaudon S, Levy R, Bonnot AM, Nguyen C, Mariolle D, Bertin F, Chabli A (2007) Drying nano particles solution on an oscillating tip at an air liquid interface: what we can learn, what we can do. Nanoscale Res Lett 2(7):309–318

72. Liu JW, Zhang SY, Qi H, Wen WC, Yu SH (2012) A general strategy for self-assembly of nanosized building blocks on liquid/liquid interfaces. Small 8(15):2412–2420
73. Pevzner A, Engel Y, Elnathan R, Ducobni T, Ben-Ishai M, Reddy K, Shpaisman N, Tsukernik A, Oksman M, Patolsky F (2010) Knocking down highly-ordered large-scale nanowire arrays. Nano Lett 10(4):1202–1208
74. Watanabe K, Kim BS, Enomoto Y, Kim IS (2011) Fabrication of uniaxially aligned poly (propylene) nanofibers via handspinning. Macromol Mater Eng 296(6):568–573
75. McAlpine MC, Ahmad H, Wang D, Heath JR (2007) Highly ordered nanowire arrays on plastic substrates for ultrasensitive flexible chemical sensors. Nat Mater 6(5):379–384
76. Li Y, Wu Y (2009) Coassembly of graphene oxide and nanowires for large-area nanowire alignment. J Am Chem Soc 131(16):5851–5857
77. Onsager L (1949) The effects of shape on the interaction of colloidal particles. Ann N Y Acad Sci 51(4):627–659
78. Fan Z, Ho JC, Jacobson ZA, Yerushalmi R, Alley RL, Razavi H, Javey A (2008) Wafer-scale assembly of highly ordered semiconductor nanowire arrays by contact printing. Nano Lett 8(1):20–25
79. Xu F, Durham JW III, Wiley BJ, Zhu Y (2011) Strain-release assembly of nanowires on stretchable substrates. ACS Nano 5(2):1556–1563
80. Yang W, Qu L, Zheng R, Liu Z, Ratinac KR, Shen L, Yu D, Yang L, Barrow JC, Ringer SP (2011) Self-assembly of gold nanowires along carbon nanotubes for ultrahigh-aspect-ratio hybrids. Chem Mater 23(11):2760–2765
81. Cisneros DA, Friedrichs J, Taubenberger A, Franz CM, Muller DJ (2007) Creating ultrathin nanoscopic collagen matrices for biological and biotechnological applications. Small 3 (6):956–963
82. Leow WW, Hwang W (2011) Epitaxially guided assembly of collagen layers on mica surfaces. Langmuir 27(17):10907–10913
83. Wen L, Wong KM, Fang Y, Wu M, Lei Y (2011) Fabrication and characterization of well-aligned, high density ZnO nanowire arrays and their realizations in Schottky device applications using a two-step approach. J Mater Chem 21(20):7090–7097
84. Ooi C, Yellen BB (2008) Field gradients can control the alignment of nanorods. Langmuir 24(16):8514–8521
85. Chen M, Sun L, Bonevich J, Reich D, Chien C, Searson P (2003) Tuning the response of magnetic suspensions. Appl Phys Lett 82(19):3310–3312
86. Gao F, Gu Z (2010) Nano-soldering of magnetically aligned three-dimensional nanowire networks. Nanotechnology 21(11):115604
87. Yuan J, Gao H, Schacher F, Xu Y, Richter R, Tremel W, Muller AH (2009) Alignment of tellurium nanorods via a magnetization-alignment-demagnetization ("MAD") process assisted by an external magnetic field. ACS Nano 3(6):1441–1450
88. Heo K, Cho E, Yang J-E, Kim M-H, Lee M, Lee BY, Kwon SG, Lee M-S, Jo M-H, Choi H-J (2008) Large-scale assembly of silicon nanowire network-based devices using conventional microfabrication facilities. Nano Lett 8(12):4523–4527
89. Hangarter CM, Rheem Y, Yoo B, Yang E-H, Myung NV (2007) Hierarchical magnetic assembly of nanowires. Nanotechnology 18(20):205305
90. Da Silva TL, Varanda LC (2011) Perpendicularly self-oriented and shape-controlled L10-FePt nanorods directly synthesized by a temperature-modulated process. Nano Res 4 (7):666–674
91. Srivastava S, Kotov NA (2008) Composite layer-by-layer (LBL) assembly with inorganic nanoparticles and nanowires. Acc Chem Res 41(12):1831–1841
92. Wang B, Ma Y, Li N, Wu Y, Li F, Chen Y (2010) Facile and scalable fabrication of well-aligned and closely packed single-walled carbon nanotube films on various substrates. Adv Mater 22(28):3067–3070
93. Li M, Bhiladvala RB, Morrow TJ, Sioss JA, Lew K-K, Redwing JM, Keating CD, Mayer TS (2008) Bottom-up assembly of large-area nanowire resonator arrays. Nat Nanotechnol 3 (2):88–92

94. Fragouli D, Buonsanti R, Bertoni G, Sangregorio C, Innocenti C, Falqui A, Gatteschi D, Cozzoli PD, Athanassiou A, Cingolani R (2010) Dynamical formation of spatially localized arrays of aligned nanowires in plastic films with magnetic anisotropy. ACS Nano 4(4): 1873–1878
95. Shekhar S, Stokes P, Khondaker SI (2011) Ultrahigh density alignment of carbon nanotube arrays by dielectrophoresis. ACS Nano 5(3):1739–1746
96. Ruda H, Shik A (2011) Principles of nanowire alignment in an electric field. J Appl Phys 109(6):064305
97. Papadakis SJ, Hoffmann JA, Deglau D, Chen A, Tyagi P, Gracias DH (2011) Quantitative analysis of parallel nanowire array assembly by dielectrophoresis. Nanoscale 3(3): 1059–1065
98. O'Riordan A, Iacopino D, Lovera P, Floyd L, Reynolds K, Redmond G (2011) Dielectrophoretic self-assembly of polarized light emitting poly (9, 9-dioctylfluorene) nanofibre arrays. Nanotechnology 22(10):105602
99. Oh K, Chung J-H, Riley JJ, Liu Y, Liu WK (2007) Fluid flow-assisted dielectrophoretic assembly of nanowires. Langmuir 23(23):11932–11940
100. Maijenburg A, Maas M, Rodijk E, Ahmed W, Kooij E, Carlen E, Blank D, Ten Elshof J (2011) Dielectrophoretic alignment of metal and metal oxide nanowires and nanotubes: A universal set of parameters for bridging prepatterned microelectrodes. J Colloid Interface Sci 355(2):486–493
101. Gates B, Mayers B, Grossman A, Xia Y (2002) A sonochemical approach to the synthesis of crystalline selenium nanowires in solutions and on solid supports. Adv Mater 14(23): 1749–1752
102. Lee TI, Choi WJ, Moon KJ, Choi JH, Kar JP, Das SN, Kim YS, Baik HK, Myoung JM (2010) Programmable direct-printing nanowire electronic components. Nano Lett 10(3): 1016–1021
103. Xu D, Shou K, Nelson BJ (2011) Dielectrophoretic assembly of carbon nanotube-based NEMS devices using floating electrodes. Microelectron Eng 88(8):2703–2706
104. Raychaudhuri S, Dayeh SA, Wang D, Yu ET (2009) Precise semiconductor nanowire placement through dielectrophoresis. Nano Lett 9(6):2260–2266
105. Fan D, Cammarata R, Chien C (2008) Precision transport and assembling of nanowires in suspension by electric fields. Appl Phys Lett 92(9):093115
106. Vijayaraghavan A, Blatt S, Weissenberger D, Oron-Carl M, Hennrich F, Gerthsen D, Hahn H, Krupke R (2007) Ultra-large-scale directed assembly of single-walled carbon nanotube devices. Nano Lett 7(6):1556–1560
107. Boote J, Critchley K, Evans S (2006) Surfactant mediated assembly of gold nanowires on surfaces. J Exp Nanosci 1(2):125–142
108. Freer EM, Grachev O, Duan X, Martin S, Stumbo DP (2010) High-yield self-limiting single-nanowire assembly with dielectrophoresis. Nat Nanotechnol 5(7):525–530
109. Bai J, Huang S, Wang L, Chen Y, Huang Y (2009) Fluid assisted assembly of one-dimensional nanoparticle array inside inorganic nanotubes. J Mater Chem 19(7): 921–923
110. Kuhn P, Puigmartí-Luis J, Imaz I, Maspoch D, Dittrich PS (2011) Controlling the length and location of in situ formed nanowires by means of microfluidic tools. Lab Chip 11(4): 753–757
111. Liu M, Chen Y, Guo Q, Li R, Sun X, Yang J (2011) Controllable positioning and alignment of silver nanowires by tunable hydrodynamic focusing. Nanotechnology 22(12):125302
112. Morrow TJ, Li M, Kim J, Mayer TS, Keating CD (2009) Programmed assembly of DNA-coated nanowire devices. Science 323(5912):352
113. Pabst W, Gregorová E, Berthold C (2006) Particle shape and suspension rheology of short-fiber systems. J Eur Ceram Soc 26(1):149–160
114. Hobbie EK, Fagan JA, Becker ML, Hudson SD, Fakhri N, Pasquali M (2008) Self-assembly of ordered nanowires in biological suspensions of single-wall carbon nanotubes. ACS Nano 3(1):189–196

115. Martin JE, Adolf D, Wilcoxon JP (1988) Viscoelasticity of near-critical gels. Phys Rev Lett 61(22):2620
116. Fry D, Langhorst B, Wang H, Becker M, Bauer B, Grulke E, Hobbie E (2006) Rheo-optical studies of carbon nanotube suspensions. J Chem Phys 124(5):054703
117. Kharchenko SB, Douglas JF, Obrzut J, Grulke EA, Migler KB (2004) Flow-induced properties of nanotube-filled polymer materials. Nat Mater 3(8):564–568
118. Seo M, Gorelikov I, Williams R, Matsuura N (2010) Microfluidic assembly of monodisperse, nanoparticle-incorporated perfluorocarbon microbubbles for medical imaging and therapy. Langmuir 26(17):13855–13860
119. Hong JS, Stavis SM, DePaoli Lacerda SH, Locascio LE, Raghavan SR, Gaitan M (2010) Microfluidic directed self-assembly of liposome–hydrogel hybrid nanoparticles. Langmuir 26(13):11581–11588
120. Ziegler KJ, Polyakov B, Kulkarni JS, Crowley TA, Ryan KM, Morris MA, Erts D, Holmes JD (2004) Conductive films of ordered nanowire arrays. J Mater Chem 14(4): 585–589
121. Zhou H, Heyer P, Kim H-J, Song J-H, Piao L, Kim S-H (2011) Reversible macroscopic alignment of Ag nanowires. Chem Mater 23(16):3622–3627
122. Yang D, Lu B, Zhao Y, Jiang X (2007) Fabrication of aligned fibrous arrays by magnetic electrospinning. Adv Mater 19(21):3702–3706
123. Su B, Wang S, Ma J, Wu Y, Chen X, Song Y, Jiang L (2012) Elaborate positioning of nanowire arrays contributed by highly adhesive superhydrophobic pillar-structured substrates. Adv Mater 24(4):559–564
124. Yu G, Li X, Lieber CM, Cao A (2008) Nanomaterial-incorporated blown bubble films for large-area, aligned nanostructures. J Mater Chem 18(7):728–734
125. Yu GH, Cao AY, Lieber CM (2007) Large-area blown bubble films of aligned nanowires and carbon nanotubes. Nat Nanotechnol 2:372
126. Lu X, Wang C, Wei Y (2009) One-dimensional composite nanomaterials: synthesis by electrospinning and their applications. Small 5(21):2349–2370
127. Li D, Wang Y, Xia Y (2004) Electrospinning nanofibers as uniaxially aligned arrays and layer-by-layer stacked films. Adv Mater 16(4):361–366
128. Greiner A, Wendorff JH (2007) Electrospinning: a fascinating method for the preparation of ultrathin fibers. Angew Chem Int Ed 46(30):5670–5703
129. He D, Hu B, Yao Q-F, Wang K, Yu S-H (2009) Large-scale synthesis of flexible free-standing SERS substrates with high sensitivity: electrospun PVA nanofibers embedded with controlled alignment of silver nanoparticles. ACS Nano 3(12):3993–4002
130. Xu C, Inai R, Kotaki M, Ramakrishna S (2004) Aligned biodegradable nanofibrous structure: a potential scaffold for blood vessel engineering. Biomaterials 25(5):877–886
131. Dersch R, Liu T, Schaper A, Greiner A, Wendorff J (2003) Electrospun nanofibers: internal structure and intrinsic orientation. J Polym Sci Part A Polym Chem 41(4):545–553
132. Katta P, Alessandro M, Ramsier R, Chase G (2004) Continuous electrospinning of aligned polymer nanofibers onto a wire drum collector. Nano Lett 4(11):2215–2218
133. Liu Y, Zhang X, Xia Y, Yang H (2010) Magnetic-field-assisted electrospinning of aligned straight and wavy polymeric nanofibers. Adv Mater 22(22):2454–2457
134. Kakade MV, Givens S, Gardner K, Lee KH, Chase DB, Rabolt JF (2007) Electric field induced orientation of polymer chains in macroscopically aligned electrospun polymer nanofibers. J Am Chem Soc 129(10):2777–2782
135. Li D, Ouyang G, McCann JT, Xia Y (2005) Collecting electrospun nanofibers with patterned electrodes. Nano Lett 5(5):913–916
136. Zhang D, Chang J (2008) Electrospinning of three-dimensional nanofibrous tubes with controllable architectures. Nano Lett 8(10):3283–3287
137. Roskov KE, Kozek KA, Wu W-C, Chhetri RK, Oldenburg AL, Spontak RJ, Tracy JB (2011) Long-range alignment of gold nanorods in electrospun polymer nano/microfibers. Langmuir 27(23):13965–13969

138. Yao Y, Gu ZZ, Zhang J, Pan C, Zhang Y, Wei H (2007) Fiber-oriented liquid crystal polarizers based on anisotropic electrospinning. Adv Mater 19(21):3707–3711
139. Law M, Sirbuly DJ, Yang PD (2007) Chemical sensing with nanowires using electrical and optical detection. Int J Nanotechnol 4(3):252–262
140. Bisht GS, Canton G, Mirsepassi A, Kulinsky L, Oh S, Dunn-Rankin D, Madou MJ (2011) Controlled continuous patterning of polymeric nanofibers on three-dimensional substrates using low-voltage near-field electrospinning. Nano Lett 11(4):1831–1837
141. Lipomi DJ, Chiechi RC, Dickey MD, Whitesides GM (2008) Fabrication of conjugated polymer nanowires by edge lithography. Nano Lett 8(7):2100–2105
142. Baji A, Mai Y-W, Wong S-C, Abtahi M, Chen P (2010) Electrospinning of polymer nanofibers: Effects on oriented morphology, structures and tensile properties. Compos Sci Technol 70(5):703–718
143. Li D, Herricks T, Xia Y (2003) Magnetic nanofibers of nickel ferrite prepared by electrospinning. Appl Phys Lett 83(22):4586–4588

Chapter 2
Synthesis of One-Dimensional Te Nanostructures

2.1 Introduction

In the recent years, one-dimensional semiconductors, such as nanowires, nanorods, and nanotubes, have received wide attention because of their promising potential applications in basic scientific research. The nanowires with diameter of 1–100 nm, a large surface area, and one-dimensional conduction have great prospects in the preparation of numerous devices. Moreover, nanowires exhibit high reactivity and can act as ideal platform to transmit microscopic particles due to their unique structural characteristics.

In general, tellurium is a spiral p-type semiconductor with a configuration of $4d^{10}\ 5s^2\ 5p^4$ and band gap energy of 0.35 eV at room temperature [1]. The crystal structure of tellurium shows highly anisotropic nature, comprising helical chains of covalently bound atoms resulting in t-Te inherent chirality, as well as a strong tendency toward 1D growth. Moreover, the peculiar properties of tellurium such as unidirectional image conductivity, nonlinear optical response, and high thermal conditions to produce piezoelectric response make it promising material for use in optoelectronic devices, gas sensors, field-effect devices, photonic crystals, radiation-cooled instruments, self-developing holographic instruments, and infrared acoustic-optic deflectors [2, 3].

Over the last several years, the several methods have been developed for the successful synthesis of Te nanocrystals in a rich variety of shapes such as zero-dimensional nanoparticles, one-dimensional nanostructures, two-dimensional nanostructures, and even more complex hierarchical structures. By using a variety of synthesis methods such as solution-phase approach, refluxing process [4], microwave-assisted synthesis in ionic liquids [5], biomolecule-assisted routes [6], visible-light-assisted technique, and vapor-phase growth [7], many Te nanostructures have been synthesized very efficiently.

J. Liu, *Well-Organized Inorganic Nanowire Films*, Springer Theses,
DOI 10.1007/978-981-10-3947-8_2

2.2 Microwave-Assisted Synthesis of Te Nanowires

The microwave chemistry has been accepted as a promising method for the liquid-phase preparation of in organic materials due to its definite advantages such as high reaction rate, low processing cost, and high yield. This rapid, efficient, and nonpolluting strategy quickly provides the homogenous solvent, uniform nucleation, and growth by the microwave irradiation of medium. It is widely used for the synthesis of a large number of nanostructures. Microwave synthesis of nanostructures has been studied by many scientists, for example, Hu and coworkers have developed the microwave-assisted ionic liquid (MAIL) method for the production of Te nanorods and nanowires [5]. Up to now, it has not been possible to synthesize high-yield, high-quality tellurium nanowires in the aqueous phase by microwave method.

With the utilization of microwave approach, Yu and coworkers have synthesized high-quality Te nanowires in water phase in high yield without using ionic liquids (Fig. 2.1a, b) [8]. The monocrystalline tellurium nanowires with a diameter of 20 nm and a length of several tens of micrometers can be synthesized by the microwave-assisted reaction by using Na_2TeO_3 as tellurium source, polyvinylpyrrolidone (PVP) as a surfactant and hydrazine hydrate as reducing agent in a microwave-sealed vessel made of quartz at 150 °C by microwave. The construction of these high-quality tellurium nanowires is extremely dependent on the reaction conditions such as the amount of PVP, the reaction time, the pH of the initial solution, and the choice of surfactants.

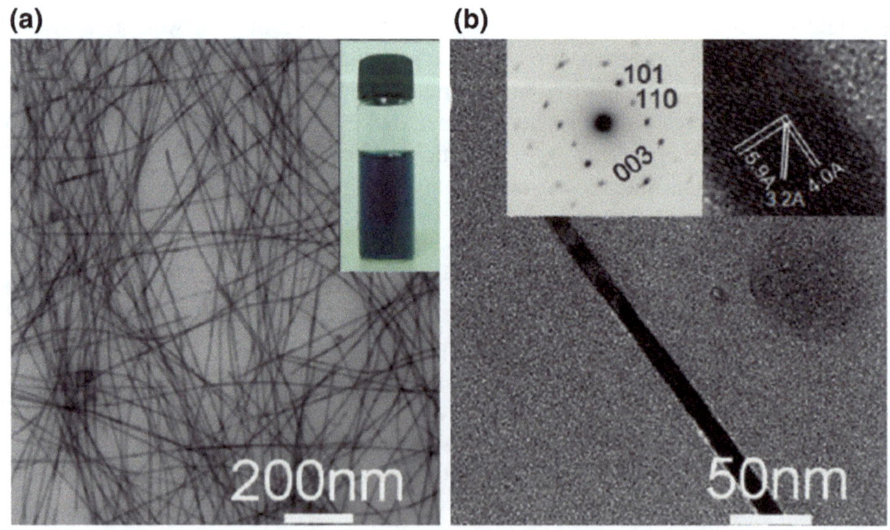

Fig. 2.1 **a** TEM image of the Te nanowires; **b** a single Te nanowire. Reprinted with permission from Ref. [8] Copyright 2010 American Chemical Society

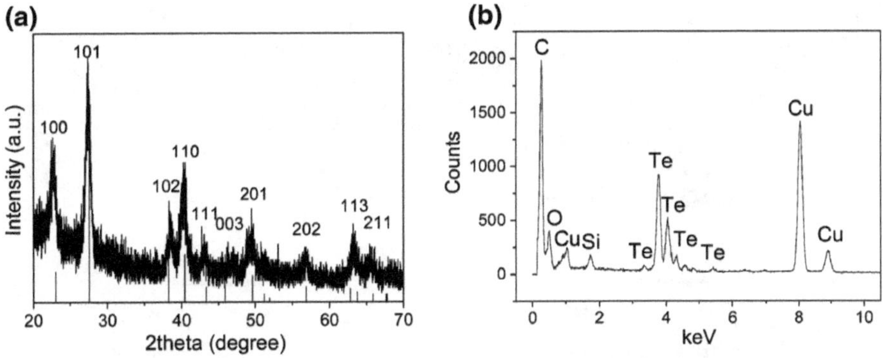

Fig. 2.2 XRD and EDS spectra of synthesized nanowires. Reprinted with permission from Ref. [8] Copyright 2010 American Chemical Society

The reaction proceeds with the reduction of Na_2TeO_3 by hydrazine. The XRD spectra presented in Fig. 2.2 shows the formation of Te nanowires with a constant hexagonal tellurium phase by a = 4.0 Å and b = 5.9 Å. Similarly, an energy-dispersive X-ray spectroscopy (EDS) spectrum shows the strong tellurium peaks, exhibiting the product as pure Tellurium.

Tellurium nanowires synthesized by microwave methods form a deep blue color solution on dispersion in water that is a characteristic color as found previously for ultrathin Te nanowires synthesized by the hydrothermal method, indicating the successful synthesis of Te nanowires by microwave method [9]. Moreover, Te nanowires show high aspect ratio with the diameter of 20 nm, and lengths up to tens of micrometers (Fig. 2.1a) with the lattice spacing of ca. 5.9, 4.0, and 3.2 Å, analogous to the lattice spacings of the (001), (100), and (101) planes for hexagonal tellurium, respectively. The plane perpendicular to the axis of the nanowire has a spacing of 0.59 nm, which is in agreement with the (001) crystal plane, showing the monocrystalline nature of tellurium nanowire (Fig. 2.1b).

2.3 Solution-Based Synthesis of Te Nanostructure

Te nanostructure with different morphologies such as nanoparticles, nanorods, and nanowires can be synthesized by using a simple solution-based approach that involves the hydrothermal treatment of a mixture containing PVP, Na_2TeO_3, NaOH, hydrazine hydrate, and ethylene glycol (EG) at 70 °C for 15 s to 3 h [10]. Moreover, the diameter of nanowires changes drastically by the addition of acetone to the reaction mixture. Figure 2.3a–e shows the SEM and TEM images of Te nanostructure indicating the change in morphology with an increase in time. Initially, at the time of 15 s, Te nanoparticles with the diameter of 7 ± 3 nm are obtained, which change to nanorods as the reaction time increases gradually. The

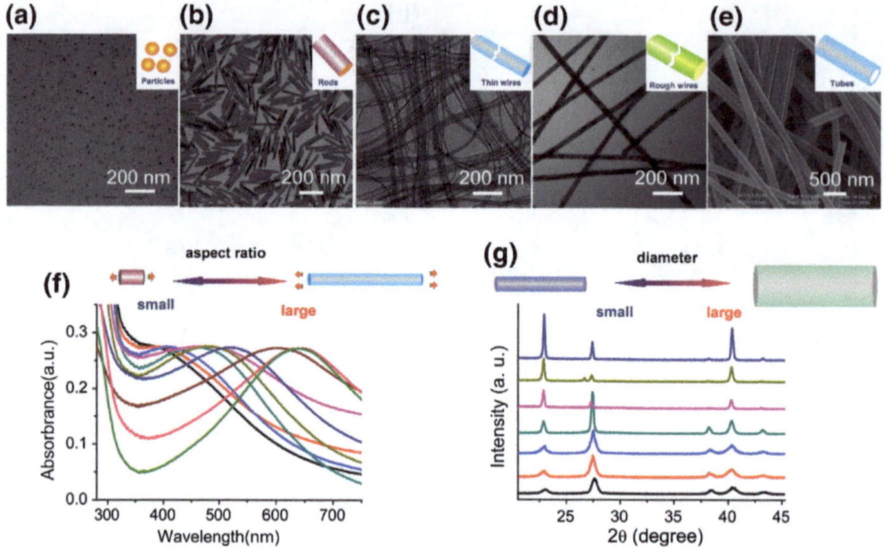

Fig. 2.3 **a–d** TEM images of Te nanostructures with different morphologies. **e** SEM image of Te nanotubes. **f** UV–Vis spectrum of Te nanostructures, from nanoparticles to nanorods with different aspect ratio. **g** XRD patterns of the as-synthesized Te nanowires with different diameters. Reprinted with the permission from Ref. [10]. Copyright 2016 WILEY-VCH

formation of Te nanorods is due to the anisotropic crystal structure of tellurium, which contains helical chains of covalently bound tellurium atoms. The chains have a strong tendency to grow along the *c*-axis direction into a one-dimensional structure [11]. Furthermore, with the increase in aspect ratio of nanorods, the absorption peak of *UV–Vis* spectra shifts from 388 to 639 nm, indicating that the UV–Vis spectra track the increase in reaction time (Fig. 2.3f). Also the diameter of nanowire changes from 7 to 150 nm with the addition of acetone and the nanotubes will be formed as the amount of acetone continues to increase (more than 50% of the total system) due to the increase in (100) proportion compared to (001) (Fig. 2.3g).

References

1. Zhuang TT, Liu Y, Li Y, Zhao Y, Wu L, Jiang J, Yu SH (2016) Integration of semiconducting sulfides for full-spectrum solar energy absorption and efficient charge separation. Angew Chem 128(22):6506–6510
2. Frank S, Poncharal P, Wang Z, de Heer WA (1998) Carbon nanotube quantum resistors. Science 280(5370):1744–1746
3. Duan X, Huang Y, Cui Y, Wang J, Lieber CM (2001) Indium phosphide nanowires as building blocks for nanoscale electronic and optoelectronic devices. Nature 409(6816):66–69

4. Yao HB, Li XB, Liu SJ, Yu SH (2009) Lamellar transition-metal molybdate-CTA mesostructured composites (metal = Ni, Co): one-pot synthesis and application in treatment of acid fuchsine. Chem Commun 44:6732–6734
5. Zhu YJ, Wang WW, Qi RJ, Hu XL (2004) Microwave-assisted synthesis of single-crystalline tellurium nanorods and nanowires in ionic liquids. Angew Chem 116(11):1434–1438
6. Lu Q, Gao F, Komarneni S (2004) Biomolecule-assisted reduction in the synthesis of single-crystalline tellurium nanowires. Adv Mater 16(18):1629–1632
7. Zhang B, Hou W, Ye X, Fu S, Xie Y (2007) 1D tellurium nanostructures: photothermally assisted morphology-controlled synthesis and applications in preparing functional nanoscale materials. Adv Funct Mater 17(3):486–492
8. Liu J-W, Chen F, Zhang M, Qi H, Zhang C-L, Yu S-H (2010) Rapid microwave-assisted synthesis of uniform ultralong Te nanowires, optical property, and chemical stability. Langmuir 26(13):11372–11377
9. Qian H-S, Yu S-H, Gong J-Y, Luo L-B, Fei L-F (2006) High-quality luminescent tellurium nanowires of several nanometers in diameter and high aspect ratio synthesized by a poly (vinyl pyrrolidone)-assisted hydrothermal process. Langmuir 22(8):3830–3835
10. Liu JW, Xu J, Hu W, Yang JL, Yu SH (2015) Systematic synthesis of tellurium nanostructures and their optical properties: from nanoparticles to nanorods, nanowires, and nanotubes. ChemNanoMat
11. Li Z, Zheng S, Zhang Y, Teng R, Huang T, Chen C, Lu G (2013) Controlled synthesis of tellurium nanowires and nanotubes via a facile, efficient, and relatively green solution phase method. J Mater Chem A 1(47):15046–15052

Chapter 3
Interface-Induced Macroscopic Nanowire Assemblies

3.1 Introduction

Within past few decades, a significant attention has been paid to one-dimensional nanomaterials such as nanowires, nanotubes, band, and especially ultrafine nanowires [1] because of their unique structure, exciting physical properties and their potential involvement in many applications [2, 3]. Over the last years, the scientific research has gradually diverted its focus from synthetic technology to orderly assembly, to create a more reasonable and orderly superstructures. The assembly of nanomaterials from a disordered state to ordered state can result in unique electrical, optical, magnetic, and spectral properties of the nanomaterials that could potentially lead to fruitful applications [4–6]. Moreover, controlling nanowire assembly results in the design of hierarchical structures with unique functionalities that play a vital role in the next generation of photovoltaic devices and other nanotechnology. Assembling nanowires to obtain macroscopic nanodevices with new functions has become an active research topic in the field of material sciences and presents a wide range of potential applications [6–11] in the traditional electronic devices, thermoelectric materials, new biochemical sensors, and optoelectronic devices. For the assembly of nanomaterials, various strategies have been developed such as Langmuir–Blodgett technique [8], external fields like electric [12] and magnetic field induced assembly [13], external nanostructures [14], and assembly induced by evaporation [15]. Liquid–liquid interface is a cost-effective self-assembly platform to assemble the nanomaterials with the interface to obtain the ordered nanowire films on the macroscopic scale. In particular, the Langmuir–Blodgett technique has appeared as an excellent way to assemble nanomaterials without hydrophobic pretreatment and also used for functional assembly of superfine and ultrafine tellurium nanowires of about 7 nm in diameter and several hundred micrometers in length. Moreover, LB technology offers an attractive assembly technology to assemble a number of nanostructures with its unique application at the air–water interface. Up to now, soft LB technique has been widely used for the assembly of

J. Liu, *Well-Organized Inorganic Nanowire Films*, Springer Theses,
DOI 10.1007/978-981-10-3947-8_3

hard nanomaterials [16] like nanoparticles [17], two-dimensional structures [18], Pt nanotubes and other 1D nanomaterials such as Ag_2Te nanowires, silver nanowires [19], Te nanowires and $W_{18}O_{49}$ nanowires [20]. This chapter focuses on the interfacial assembly of nanowires and brief overview will be conducted to explain how the interface can be used as an assembly platform to fabricate nanowire assemblies.

3.2 Interface-Induced Assembly of One-Dimensional Nanomaterials

3.2.1 Langmuir–Blodgett Technology for Assembly of Nanowires

Irving Langmuir and Katherine Blodgett revealed the science of Langmuir–Blodgett films by examining the transfer of Langmuir monolayers onto substrates in early twentieth century. This technique is considered as the most conventional strategy for the well-organized building of moieties in one monolayer at a time for nanofilm fabrication. With the utilization of LB apparatus, the monolayer is constructed at the gas–liquid interface that comprises a Langmuir trough to lower or raise the substrate through the gas–liquid interface and a surface pressure sensor that controls the movable barrier [21] and an automated movable barrier, which moves during the deposition process in order to maintain a controlled surface pressure (Fig. 3.1). Furthermore, the film formed by this method at the air–water interface can be easily transferred to the substrate of choice [22] without any damage to the film.

3.2.1.1 LB Assembly of UltraFine and UltraLong Te Nanowires

Ultrathin hydrophilic Te nanowires having an aspect ratio of 104 can be aligned over a large area to form well-defined periodic mesostructures by Langmuir–Blodgett technique. The assembly process starts by the filling of Langmuir–Blodgett trough with distilled water over the top by about 2 mm and then adding

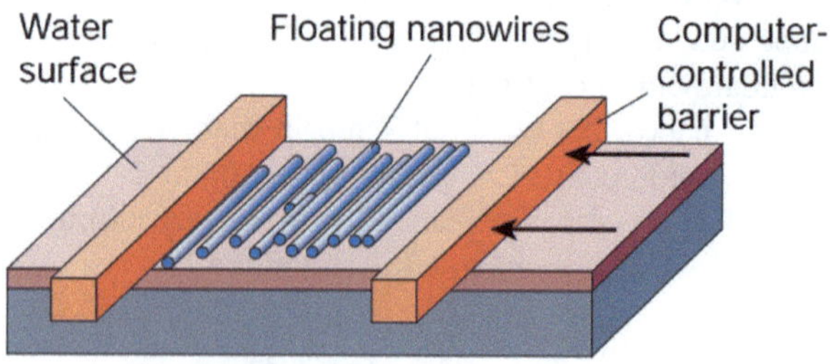

Fig. 3.1 Graphical design for LB nanowire assembly process. Reprinted with the permission from Ref. [23]. Copyright 2003 Nature Publishing Group

the nanowire solution on water reservoir. Moreover, in this process, there is no need for extra hydrophobic pretreatment or functionalization after synthesis [8]. The centrifuged nanowires are first dispersed in *N,N*-dimethylformamide (DMF) at room temperature to avoid the hydrophobization and then further added to a mixture of DMF and chloroform mixture before adding to the water–air interface. The high amount of DMF and chloroform prevents the good dispersion and uniformity of nanowires at the outset solution, respectively. So the ratio of dispersing solvent (DMF: $CHCl_3$) plays a very important role in the assembly of nanowires.

The assembly process highly depends on various factors such as solvent, surface pressure, and time as discussed below. For aligning the nanowires, good dispersion and evaporation are key parameters that depend on the choice and amount of solvent. In the case of the high amount of DMF, precipitation of Te nanowires occurs at the bottom of LB trough that hampers the dispersion of the nanowires at the air–water interface, whereas the high amount of $CHCl_3$ decreases the homogeneity of the solution.

After the evaporation of the mixed solvent, some parallel nanowires of the ordered portion are randomly oriented at the air–water interface due to the interfacial tension. The interface nanowires shrink and produce a compact structure [24] that becomes tidy as the surface tension increases. The π-A isotherms of the surface tension of Te nanowires are presented in Fig. 3.2a, indicating 4 different stages. Initially, the pressure is independent of barriers movement and the nanowires come closer to each other under the effect of van der Waals forces and capillary forces. In the second stage, the formation of concentrated Te nanowires monolayer occurs due to decrease in area and increase in surface tension. The third stage shows constant pressure with the decrease in area and the loosely arranged nanowires become compact. The slope of the π-A curve directly reveals the rate of change of the surface pressure from 0 to 23 mN/m, which indirectly shows the behavior of the nanowires (Fig. 3.2a, b).

The nanowires are arranged in an irreversible manner parallel to the pusher at the pressure of 23 mN/m for 10 h. Actually, the ordering of the monolayer increases due to the reduction of entropy. The nanowire alignment for different time 1, 3, 5, and 9 h between stages (III) and (IV) are highlighted in Fig. 3.2c, d show the increase in nanowire alignment with time. Moreover, the monolayer will be substantially damaged for the time higher than 10 h due to a decrease in entropy. On the substrate, Te nanowires are parallel and oblique as they are long and flexible and wrapped with PVP Fig. 3.3a, b. Moreover, the lattice spacing is consistent with the (001), (100), and (101) planes of hexagonal tellurium and the angle between the (001) plane and the (101) plane is 55.4° that matches well with the calculated structure of the crystal structure Fig. 3.3c [8].

Furthermore, the LB technique has also been effectively introduced for the formation of nanomesh that can be obtained under harsh conditions such as high temperature and on a special substrate surface [25, 26]. Figure 3.4a illustrates the

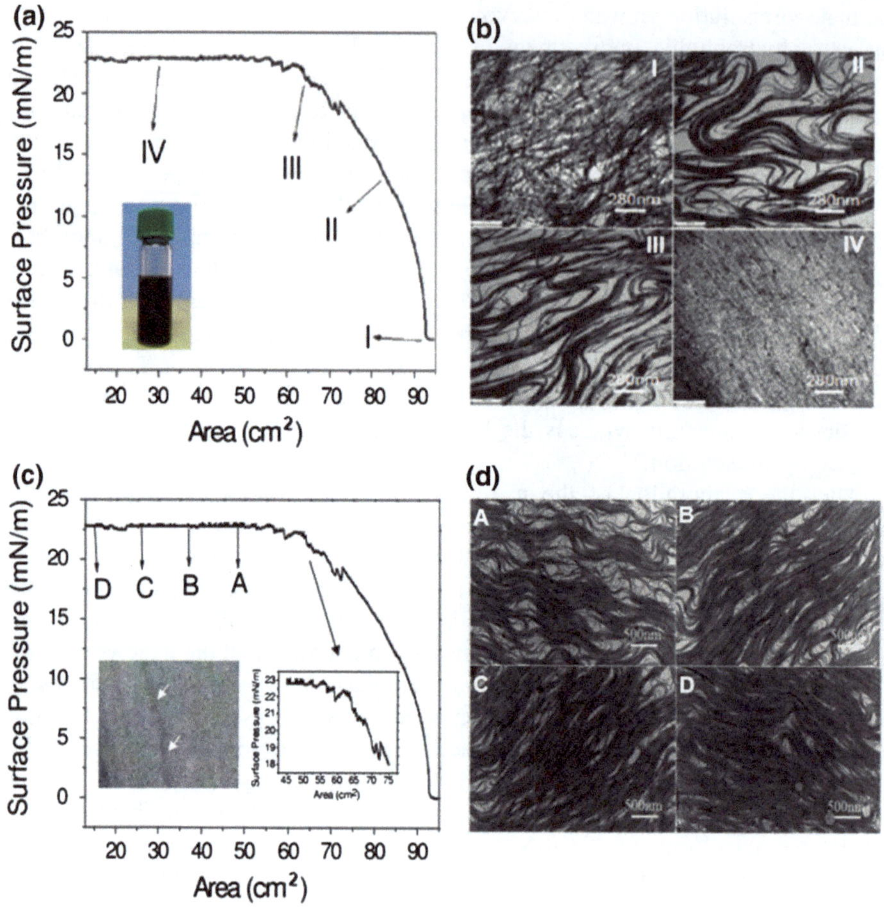

Fig. 3.2 **a** π-A isotherms, **b** TEM images of the Te nanowire, **c** π-A isotherms of the Te nanowire collected between the stages (III) and (IV), **d** corresponding TEM images of the Te nanowires. Reprinted with the permission from Ref. [8]. Copyright 2010 American Chemical Society

schematic design for transfer of two layers of aligned nanowires into mesh-like structures with designed crossing angles and can be designed into nanomesh-like crossed patterns with a tunable crossed angles such as 45°, 60°, 65°, 70°, and 85°, respectively, indicating that the crossing angle between the two layers of nanowires increases with the tuning angle [8].

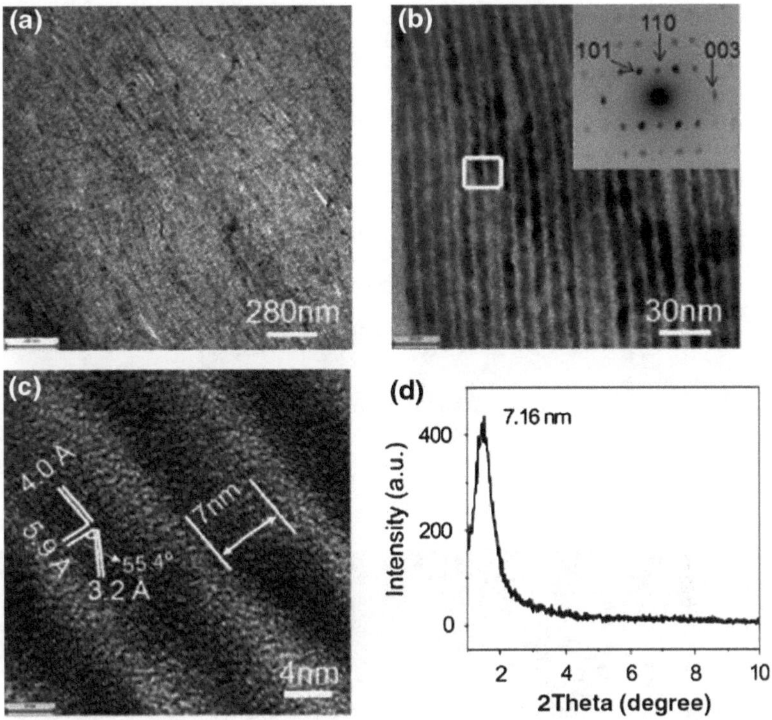

Fig. 3.3 **a**, **b** (TEM) images of the monolayer assembly of Te nanowires, **c** (HRTEM) image of the aligned Te nanowires, **d** SAXRD pattern of aligned Te nanowires. Reprinted with the permission from Ref. [8]. Copyright 2010 American Chemical Society

3.2.1.2 Ordered Ultrathin Telluride Nanowires and Hetero-nanowire Film

Tellurium and telluride are considered well-known semiconductor due to very narrow bandgap and many interesting properties such as nonlinear optical response, light guide, piezoelectric response, and high pyroelectricity, which make them valuable in many applications such as optoelectronic devices, photonic crystal field-effect devices, gas sensors, field-effect transistors, topological insulators, and holographic recording [27–30]. By using well-defined ultrathin single crystalline TeNW patterns as templates, macroscale ordered ultrathin telluride nanowire (NW) films (Ag_2Te, Cu_2Te, PbTe), and tellurium/telluride hetero-nanowire films (Te-Ag_2Te, Cu_2Te-Te-Ag_2Te-PbTe) can be rapidly fabricated through chemical transformation reactions. In order to obtain different telluride films, initially, tellurium nanowires are assembled by LB technology to attain an extensive range of ordered structures. By immersing the tellurium template in the silver nitrate solution, a color change occurs from dark blue to brown, indicating the formation of silver telluride nanowires (Fig. 3.5a, e). The degree of the order of silver telluride is

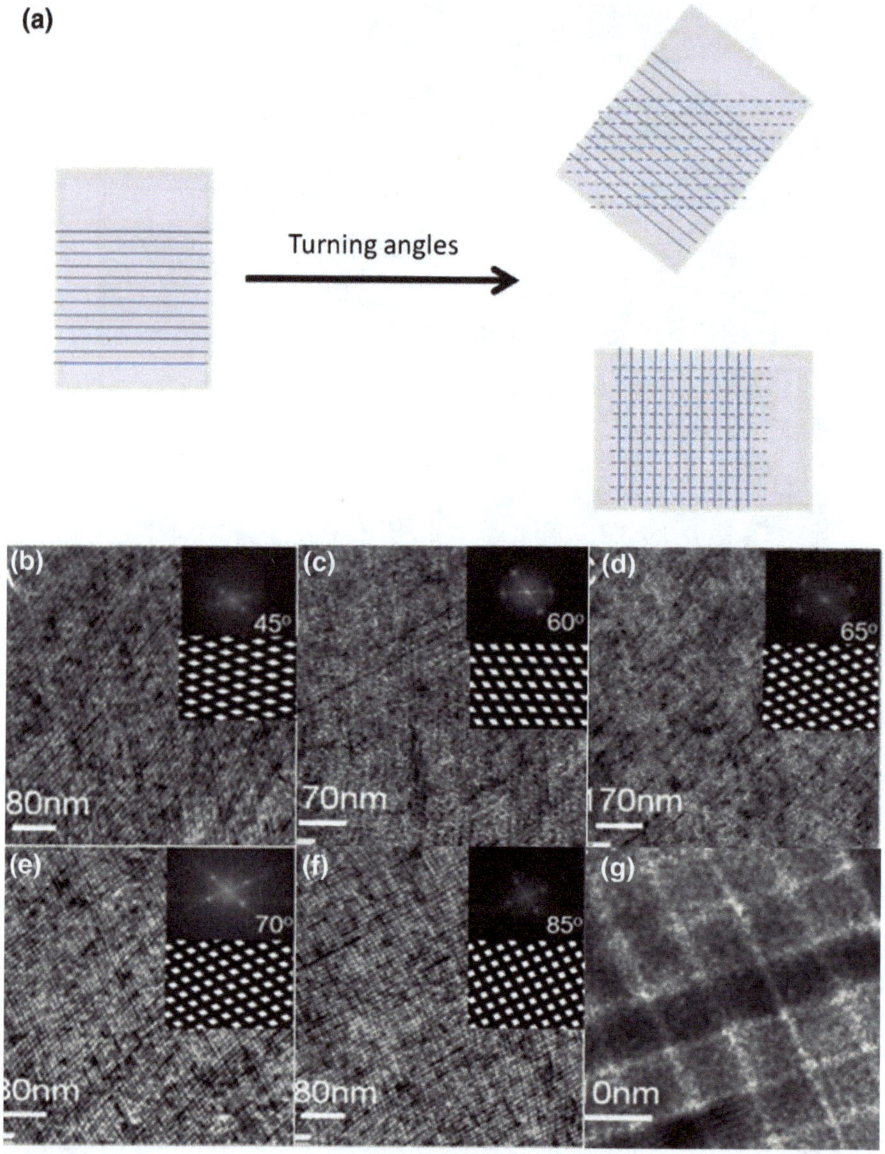

Fig. 3.4 **a** Illustration of the steps for the assembly of Te nanowire film mesostructures. **b–g** TEM micrographs of the nanomesh-like assemblies designed by packing two layers of Te nanowire films with different crossed angles. Reprinted with the permission from Ref. [8]. Copyright 2010 American Chemical Society

strongly dependent on the order of the tellurium nanowires, that is, on the transformation of the single-layer tellurium nanowire into monolayer silver telluride nanowire films, the period becomes 8.9 nm from 6.8 nm, indicating the variation of

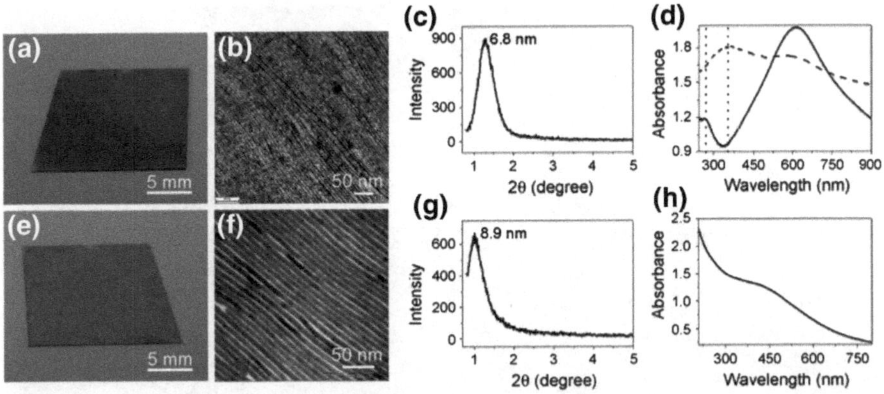

Fig. 3.5 **a, e** Photographs of ordered Te and Ag_2TeNW monolayers, **b, f** TEM images of Te and Ag_2TeNW monolayers, **c, g** SAXRD spectrums of ordered Te and Ag_2TeNW monolayers, **d** UV–Vis spectrum of the aqueous solution and ordered Te nanowires, **h** UV–Vis spectrum of an Ag_2TeNW monolayer. Reprinted with permission from Ref. [31]. Copyright 2012 WILEY-VCH

the period with the diameters of the TeNWs Fig. 3.5c, g. The UV study shows the two absorption peaks located at 586 nm and 450 nm, respectively, showing an optical change due to the pattern transfer process (Fig. 3.5d, h). Moreover, a red shift appears in the surface plasmon resonance while comparing the UV–VIS spectrum of the ordered TeNW monolayer on a quartz substrate and TeNWs in aqueous solution, indicating an increase in order [31].

This chemical transformation process can be extended from silver telluride nanowires film to other telluride nanowires films such as zinc telluride nanowire films by ion-conversion and chromium telluride by using silver telluride nanowires film as a template. Similarly, chromium telluride nanowire templates can be further converted to platinum-telluride nanotubes [31].

Moreover, mesoscopic structure of these hetero-structure nanowires can be designed with the help of LB strategy that can be attained by replacing Te nanowires with Ag_2Te nanowires or Pt nanotubes (Fig. 3.6) with the vertical placement of Ag_2Te or Pt nanotube monolayer on the first layer, resulting a staggered Ag_2Te nanowire structure (Fig. 3.6c, d) [31]. In conclusion, these chemically active Te nanowires can be used not only as a reductant, but also as a template to directly achieve ordered structures of other inorganic one-dimensional nanowires via assembly. Furthermore, this template conversion method provides us a new strategy of chemical conversion to obtain ordered nanowire films, as well as their hetero-structures [31].

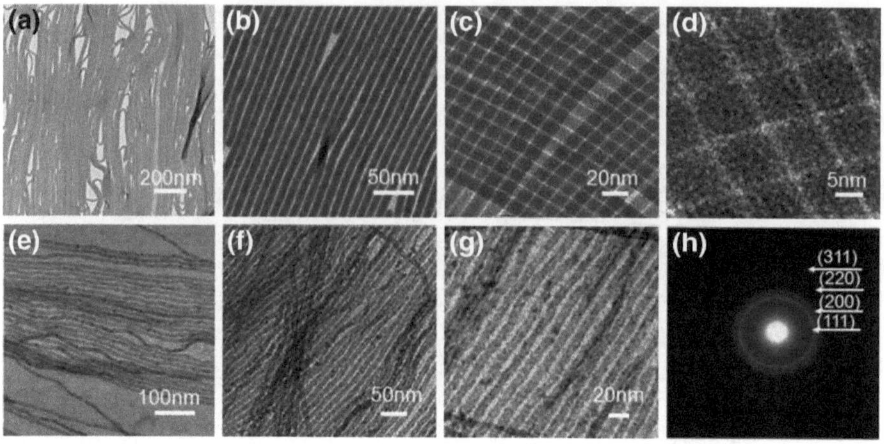

Fig. 3.6 **a–d** TEM photographs of Ag_2Te assembly. **e–g** TEM images of assembled Pt nanotubes; **c**, **d** crossed layers of Ag_2Te nanowires; **e–g** TEM images of a monolayer of Pt nanotubes; **h** SAED pattern taken on the Pt nanotubes presented in (**g**). Reprinted with permission from Ref. [31]. Copyright 2012 WILEY-VCH

3.2.1.3 Co-assembly of Ag and Te Nanowires

Nanowires assembly at microscale offers an extensive variety of potential applications with low cost and improved performance. Moreover, to prepare nanowires film in wafer scale, rational assembly strategies has been employed for novel collective properties and device fabrication at a practical scale [6, 7, 32–34]. Nanowire film assembly and nanofilm gap can be controlled by using different nanometer reactive nanowires. In order to fabricate the film of two different nanowires (Ag and Te), LB technique can be used to obtain a large area of ordered nanowire films. Te and Ag nanowires are mixed in a solvent mixture of *N,N*-dimethylformamide (DMF) and chloroform (CHCl3) to form a homogeneous single-phase solution at room temperature [35]. After that, the solution is dispersed onto a surface of deionized water with a syringe in order to form a monolayer. The nanowire network is obtained by the deposition of two layers into a mesh-like structure with the crossing angle. Furthermore, by the etching process, Te nanowires of the co-assemblies can be removed, leaving an Ag nanowire network with pitch tuned by the Te nanowires (Fig. 3.7).

3.2.1.4 LB Assembly of Ultrathin $W_{18}O_{49}$ Nanowires

Semiconductor nanowires due to their distinctive physical and chemical properties are considered effective candidates in various applications [10, 36–39]. During the past few years, various strategies have appeared to assemble semiconductor nanowires into complex and novel mesostructures, producing new functionalities

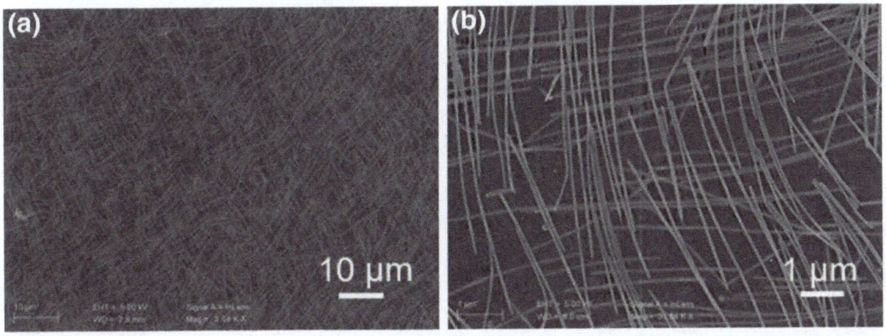

Fig. 3.7 **a** Low and **b** high magnification image of Ag nanowires assembly. Reprinted with permission from Ref. [35]. Copyright 2014 WILEY-VCH

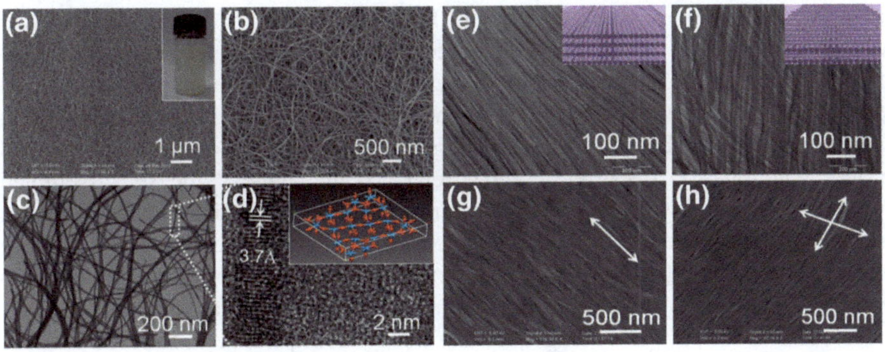

Fig. 3.8 **a–d** Morphology and structure of the as-obtained $W_{18}O_{49}$ nanowires. **e–h** Two styles (parallel and cross) of ordered nanowire nanostructures. Reprinted with permission from Ref. [50]. Copyright 2013 American Chemical Society

and properties, which make them applicable in a wide range of traditional electronic devices (logic and memory) to novel chemical and biomolecular sensors, opto-electronic devices, and thermoelectric materials [6, 7, 10]. In traditional semiconductors, tungsten oxide is considered a significant n-type semiconductor due to its photocatalytic, gas sensor, and electrochromic applications [40–45]. Electrochromic phenomenon which is related to color change induced in selected materials by a reversible electrochemical process develops due to the production of different electronic absorption bands on the visible region [46, 47]. This process is reversible and interrelated to redox states switching. In the family of well-known electrochromic materials, tungsten oxide presents a colorless and blue alternating change with the applied positive and negative voltage changes. This phenomenon was discovered by Deb in 1969 and has been extensively studied [48, 49].

The well-organized monolayer of tungsten oxide nanowires with periodic structures can be fabricated by Langmuir–Blodgett (LB) technique [50]. The surfactant PVP plays an important role in the assembly of nanowires [8, 34] and has been used as a kind of stabilizer or coating agent in the synthesis process and also plays a role in preventing the aggregation of product synthesis. In the absence of surfactant, it is difficult to assemble Tungsten oxide nanowires but with the coating of PVP, a well-ordered array of the tungsten oxide nanowires can be achieved by LB technology (Fig. 3.8).

3.2.2 Assembly of Nanowires at Liquid–Liquid Interface

Recently, research on nanomaterial self-assembly has gained much attention due to their potential in optics, magnetism, and electronics. Their applications are promising because of their significant differences compared with irregular and disordered nanomaterials [51–55]. Various strategies have been developed, including convection assemblies [56], LB techniques [7, 8, 57], layer-layer assemblies [54, 55], external driving forces (e.g., magnetic, electric, or shearing) [58, 59] and in situ methods to form the interface where the solvent evaporates [9, 10, 31]. However, the above-mentioned methods have some limitations in a wide range of applications, such as long processing time, low throughput, and proper functionality, because of the special arrangements of the self-assembled bodies. For assembly of multiple building blocks, the liquid interface offers an excellent platform on the basis of interfacial-ordering effect, attracting more attention nowadays [60–63]. Numerous methods like switching the ligand for a contact angle of 90° of the particles with the water/oil interface [64–65], adding other solvents [66, 67] and using external forces [68, 69] can be introduced to assist the interfacial assembly. Due to the general applicability and simple operation of this technique, the liquid/liquid interface offers versatile opportunities to assemble many kinds of nanosized building blocks.

Yu et al. developed an assembly method by the formation of a family of interfaces with two immiscible liquids on which many nanosized building blocks with altered shapes and dimensionalities can be fabricated at room temperature by the construction of large-area self-assembled monolayer or multilayer nanofilms [70]. This strategy integrated initially by the assembly of many nanosized building blocks with various dimensions and shapes, including zero-dimensional nanoparticles (NPs), one-dimensional nanowires (NWs), two-dimensional nanosheets (NSs), and nanocubes (NCs) and then extended to significant interfacial systems containing different immiscible liquids and external solvents by sweeping the limits of the particular interface [70].

The schematic illustration of the formation of the self-assembled nanowire is presented in Fig. 3.9. Initially, the nanomaterials such as silver nanowires are dispersed in a solution named as Solution A, which is referred to a mother liquor containing suspension of nanocubes, silver nanowires etc. Whereas, solution B is an almost completely hydrophobic organic liquid such as chloroform, dichloro-dichloroethane, methane, carbon tetrachloride, chlorobenzene, and bromobenzene.

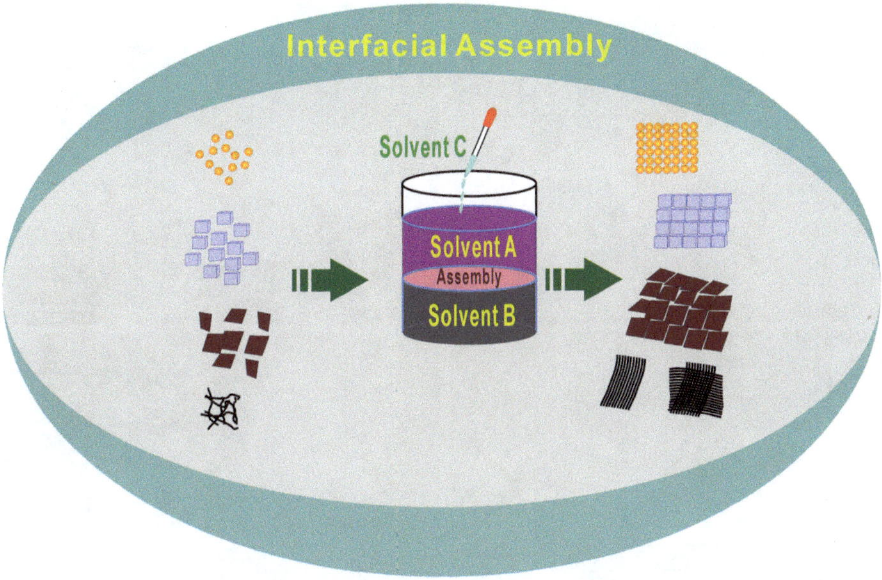

Fig. 3.9 Design of liquid/liquid interfacial assembly of different nanosized building blocks: Reprinted with permission from Ref. [69]. Copyright 2012 WILEY-VCH

For the formation of ordered nanofilms, the initial step is to drop the appropriate amount of solvent C, which provides the phenomena of nanobrick aggregate force. A small amount of solvent C cannot break the interface equilibrium, which hinders the arrangement by interfacial-ordering effects, whereas in the case of the high amount of solvent C, it is difficult to obtain a homogeneous film at a liquid/liquid interface. After the dispersion, the film appears at the oil/water interface, indicating the successful assembly of nanosized building blocks by this process. Moreover, the area of the film formed between the liquid and liquid interfaces can be several centimeters large, determined by the size of the container.

By this technique, silver nanowires having high aspect ratio are assembled to a close-packed structure. At the water/chloroform interface, macroscopic scale films of Ag NW assemblies can be quickly fabricated after adding a suitable amount of acetone as solvent C (5 mL of chloroform as solvent B is covered with 5 mL of aqueous solution of Ag NWs as solution A and 1 mL of acetone as solvent C). After the assembly process at the interface, the films of well-organized Ag NWs can be transferred to many substrates, including planar or uneven substrates (Fig. 3.10) [70].

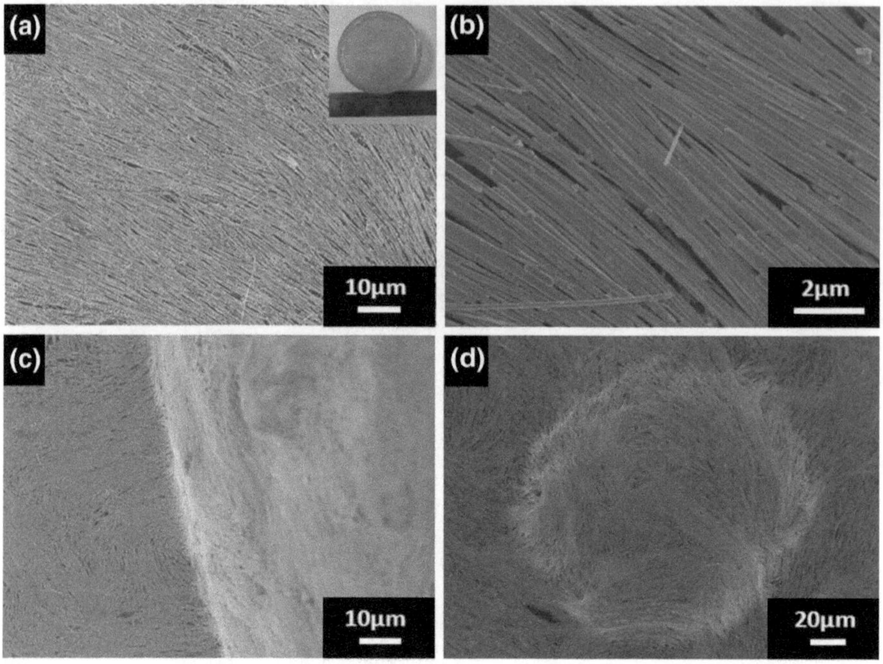

Fig. 3.10 SEM images of Ag NW films formed at the interface of an aqueous suspension of Ag NWs and chloroform with the addition of acetone. Reprinted with permission from Ref. [69]. Copyright 2012 WILEY-VCH

3.2.3 Shear Force-Induced Assembly of Silver and Tellurium Nanowires

Recently, with substantial advancement in 1-D nanomaterials, an extensive series of assembly strategies have been [38, 71] established to create and fabricate functional well-defined ordered superstructures or complex structural design with desired properties [6, 10, 72, 73]. For the assembly of nanowires, a variety of strategies has been developed but still major efforts are needed for further development in nanowires assembly strategies to expand their potential applications [74–76]. Recently, a new method based on the capillary flow of nanowires was developed for ordering hydrophilic high aspect ratio nanowires (Ag, Te) (Fig. 3.11) [77]. The formation process is strongly dependent on various flow parameters such as flow rate, the concentration of Ag nanowire suspensions and flow distance. In a typical experiment, a capillary is connected with a 10 mL syringe using polytetrafluoroethylene (PTFE) capillary as a bridge and filled with nanowire suspension. Afterward, at the flow rate of 800 $\mu L\ min^{-1}$, the nanowires suspension in the syringe is introduced into capillary with the further addition of ethanol to flow through the capillary with the same flow rate and then kept for drying.

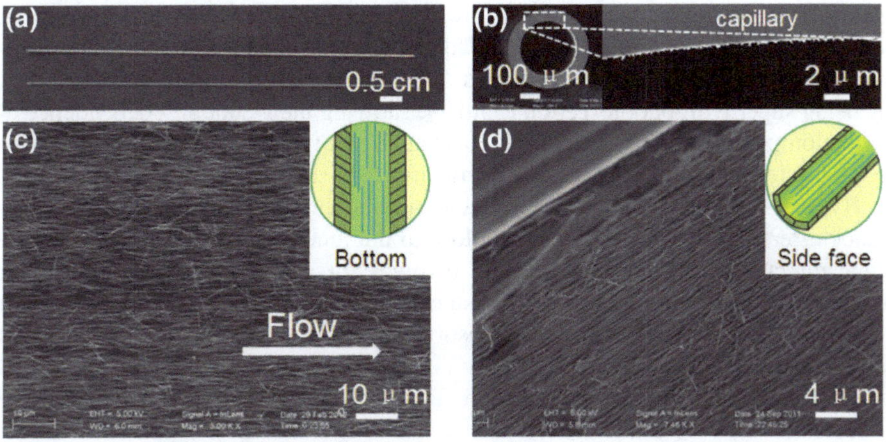

Fig. 3.11 SEM images of Ag and Te nanowire assemblies in the glass capillary; Reprinted with the permission of [77]. Copyright 2013 Nature Publishing Group

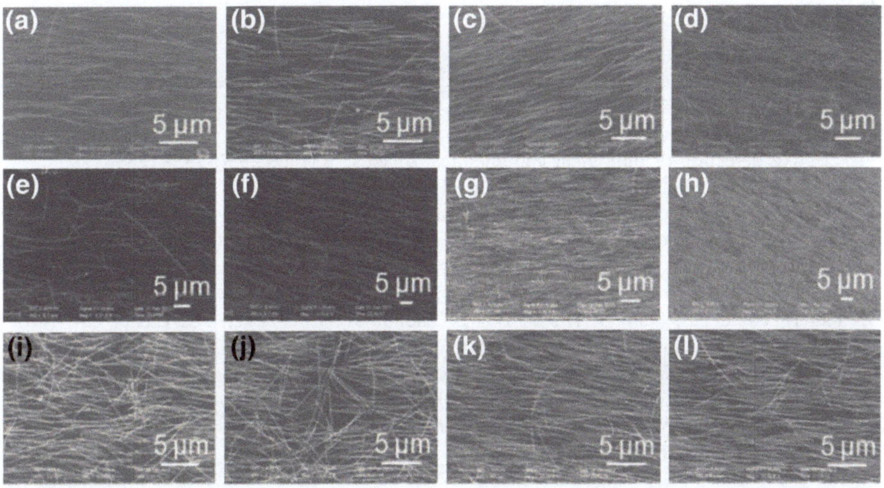

Fig. 3.12 **a–d** Influence of the flow rate of the assembly of Ag nanowires, **e–h** Influence of the concentration of the nanowire suspension on the assembly of Ag nanowires, **i–l** Influence of the flow distance on the assembly of Ag nanowires; Reprinted with permission from Ref. [77]. Copyright 2012 Nature Publishing Group

The formation of silver nanowires film is strongly affected by various factors such as the flow rate of the nanowire solution, the hydrophilicity of the capillary, the flow distance, and the concentration. While considering the flow rate of nanowires, Fig. 3.12a–d shows that the degree of nanowire alignment increases with increase in flow rate from 200 to 400 and 800 μL min^{-1}, but flow rate higher than 800 μL min^{-1} results in a decrease in the degree of nanowire ordering. Hence, higher flow rate creates larger sheer forces, leading to better alignment. However, too high flow rate decreases the sheer forces between the capillary and the nanowires. Furthermore, no nanowire film is observed on the inner wall of the capillary by using superhydrophobic Teflon capillary, showing the wettability dependence of formation process. Similarly, a high concentration of nanowires and distance also helps the formation of well-aligned nanowires film (Fig. 3.12i–l).

By this approach, the film formation occurs through the stacking of silver nanoclusters tightly in one piece and close to the inner wall of the capillary in the same direction. The major interaction involves for assembly process is van der Waals forces between the inner wall of the capillary and the PVP on the surface of the silver nanowires that allow the absorption and assembly of one-dimensional silver nanowires along the axial direction on the inner wall of the capillary without any pretreatment and functionalization. Remarkably, this technique can also be used to assemble other nanowires, like ultrathin Te nanowires with diverse aspect ratios.

References

1. Cademartiri L, Ozin GA (2009) Ultrathin nanowires—a materials chemistry perspective. Adv Mat 21(9):1013–1020
2. Cui Y, Wei Q, Park H, Lieber CM (2001) Nanowire nanosensors for highly sensitive and selective detection of biological and chemical species. Science 293(5533):1289–1292
3. Yuan J, Liu X, Akbulut O, Hu J, Suib SL, Kong J, Stellacci F (2008) Superwetting nanowire membranes for selective absorption. Nat Nanotech 3(6):332–336
4. Yaman M, Khudiyev T, Ozgur E, Kanik M, Aktas O, Ozgur EO, Deniz H, Korkut E, Bayindir M (2011) Arrays of indefinitely long uniform nanowires and nanotubes. Nat Mat 10(7):494–501
5. Yan H, Choe HS, Nam S, Hu Y, Das S, Klemic JF, Ellenbogen JC, Lieber CM (2011) Programmable nanowire circuits for nanoprocessors. Nature 470(7333):240–244
6. Liu J-W, Liang H-W, Yu S-H (2012) Macroscopic-scale assembled nanowire thin films and their functionalities. Chem Rev 112(8):4770–4799
7. Tao AR, Huang J, Yang P (2008) Langmuir-Blodgettry of nanocrystals and nanowires. Acc Chem Res 41(12):1662–1673
8. Liu J-W, Zhu J-H, Zhang C-L, Liang H-W, Yu S-H (2010) Mesostructured assemblies of ultrathin superlong tellurium nanowires and their photoconductivity. J Am Chem Soc 132(26):8945–8952
9. Bai C, Liu M (2012) Implantation of nanomaterials and nanostructures on surface and their applications. Nano Today 7(4):258–281
10. Yan H, Choe HS, Nam S, Hu Y, Das S, Klemic JF, Ellenbogen JC, Lieber CM (2011) Programmable nanowire circuits for nanoprocessors. Nature 470(7333):240–244
11. Liu JW, Xu J, Liang HW, Wang K, Yu SH (2012) Macroscale ordered ultrathin telluride nanowire films, and tellurium/telluride hetero-nanowire films. Angew Chem Int Ed 51(30): 7420–7425
12. Smith PA, Nordquist CD, Jackson TN, Mayer TS, Martin BR, Mbindyo J, Mallouk TE (2000) Electric-field assisted assembly and alignment of metallic nanowires. Appl Phys Lett 77(9):1399

13. Sheparovych R, Sahoo Y, Motornov M, Wang S, Luo H, Prasad PN, Sokolov I, Minko S (2006) Polyelectrolyte stabilized nanowires from Fe3O4 nanoparticles via magnetic field induced self-assembly. Chem Mat 18(3):591–593
14. Mann S (2009) Self-assembly and transformation of hybrid nano-objects and nanostructures under equilibrium and non-equilibrium conditions. Nat Mater 8(10):781–792
15. Moon GD, Lee TI, Kim B, Chae G, Kim J, Kim S, Myoung J-M, Jeong U (2011) Assembled monolayers of hydrophilic particles on water surfaces. ACS Nano 5(11):8600–8612
16. Acharya S, Panda AB, Belman N, Efrima S, Golan Y (2006) A semiconductor-nanowire assembly of ultrahigh junction density by the Langmuir-Blodgett technique. Adv Mat 18 (2):210–213
17. Guo Q, Teng X, Rahman S, Yang H (2003) Patterned Langmuir-Blodgett films of monodisperse nanoparticles of iron oxide using soft lithography. J Am Chem Soc 125(3): 630–631
18. Cote LJ, Kim F, Huang J (2008) Langmuir-Blodgett assembly of graphite oxide single layers. J Am Chem Soc 131(3):1043–1049
19. Tao A, Kim F, Hess C, Goldberger J, He R, Sun Y, Xia Y, Yang P (2003) Langmuir-Blodgett silver nanowire monolayers for molecular sensing using surface-enhanced Raman spectroscopy. Nano Lett 3(9):1229–1233
20. Liu BJ-W, Zheng J, Wang J-L, Xu J, Li H-H, Yu S-H (2013) Ultrathin W18O49 nanowire assemblies for electrochromic devices. Nano Lett 13(8):3589–3593
21. Evans DF, Wennerstrom H (1999) Colloidal domain. Wiley-Vch
22. Blodgett KB, Langmuir I (1937) Built-up films of barium stearate and their optical properties. Phy Rev 51(11):964
23. Yang P (2003) Nanotechnology: wires on water. Nature 425(6955):243–244
24. Li YJ, Huang WJ, Sun SG (2006) A universal approach for the self-assembly of hydrophilic nanoparticles into ordered monolayer films at a toluene/water interface. Angew Chem Int Ed 45(16):2537–2539
25. Corso M, Auwärter W, Muntwiler M, Tamai A, Greber T, Osterwalder J (2004) Boron nitride nanomesh. Science 303(5655):217–220
26. Trembulowicz A, Ciszewski A (2007) A square titanium nanomesh on W (100). Nanotechnology 18(34):345303
27. Wang Y, Tang Z, Podsiadlo P, Elkasabi Y, Lahann J, Kotov NA (2006) Mirror-like photoconductive layer-by-layer thin films of te nanowires: the fusion of semiconductor, metal, and insulator properties. Adv Mat 18(4):518–522
28. Kong D, Cui Y (2011) Opportunities in chemistry and materials science for topological insulators and their nanostructures. Nat Chem 3(11):845–849
29. Kong D, Chen Y, Cha JJ, Zhang Q, Analytis JG, Lai K, Liu Z, Hong SS, Koski KJ, Mo S-K (2011) Ambipolar field effect in the ternary topological insulator $(BixSb_{1-x})$ $2Te_3$ by composition tuning. Nat Nanotech 6(11):705–709
30. Tangney P, Fahy S (2002) Density-functional theory approach to ultrafast laser excitation of semiconductors: application to the A1 phonon in tellurium. Phy Rev B 65(5):054302
31. Liu JW, Xu J, Liang HW, Wang K, Yu SH (2012) Macroscale ordered ultrathin telluride nanowire films, and tellurium/telluride hetero-nanowire films. Angew Chem Int Ed 51(30): 7420–7425
32. Yu G, Lieber CM (2010) Assembly and integration of semiconductor nanowires for functional nanosystems. Pure App Chem 82(12):2295–2314
33. Wang MC, Gates BD (2009) Directed assembly of nanowires. Mater Today 12(5):34–43
34. Liu J-W, Wang J-L, Huang W-R, Yu L, Ren X-F, Wen W-C, Yu S-H (2012) Ordering Ag nanowire arrays by a glass capillary: a portable, reusable and durable SERS substrate. Sci Rep 2
35. Liu JW, Wang JL, Wang ZH, Huang WR, Yu SH (2014) Manipulating nanowire assembly for flexible transparent electrodes. Angew Chem Int Ed 53(49):13477–13482
36. Law M, Goldberger J, Yang P (2004) Semiconductor nanowires and nanotubes. Ann Rev Mater Res 34:83–122

37. Thomson JW, Lawson G, O'Brien P, Klenkler R, Helander MG, Petrov S, Lu ZH, Kherani NP, Adronov A, Ozin G (2010) Flash nano-welding: investigation and control of the photothermal response of ultrathin bismuth sulfide nanowire films. Adv Mater 22(39): 4395–4400
38. Hochbaum AI, Yang P (2009) Semiconductor nanowires for energy conversion. Chem Rev 110(1):527–546
39. Xu S, Qin Y, Xu C, Wei Y, Yang R, Wang ZL (2010) Self-powered nanowire devices. Nat Nanotech 5(5):366–373
40. Moshofsky B, Mokari T (2012) Length and diameter control of ultrathin nanowires of substoichiometric tungsten oxide with insights into the growth mechanism. Chem Mat 25(8): 1384–1391
41. Zhang J, Tu J-P, Xia X-H, Wang X-L, Gu C-D (2011) Hydrothermally synthesized WO_3 nanowire arrays with highly improved electrochromic performance. J Mater Chem 21(14): 5492–5498
42. Lu X, Zhai T, Zhang X, Shen Y, Yuan L, Hu B, Gong L, Chen J, Gao Y, Zhou J (2012) WO_{3-x}@Au@MnO_2 core-shell nanowires on carbon fabric for high-performance flexible supercapacitors. Adv. Mater. 24(7):938–944
43. Li X-L, Lou T-J, Sun X-M, Li Y-D (2004) Highly sensitive WO_3 hollow-sphere gas sensors. Inorg Chem 43(17):5442–5449
44. Li X, Li F, Yang C, Ge W (2001) Photocatalytic activity of WO_x-TiO_2 under visible light irradiation. J Photochem Photobiol 141(2):209–217
45. Liao C-C, Chen F-R, Kai J-J (2006) WO_{3-x} nanowires based electrochromic devices. Sol Energ Mat Sol 90(7):1147–1155
46. Whang D, Jin S, Wu Y, Lieber CM (2003) Large-scale hierarchical organization of nanowire arrays for integrated nanosystems. Nano Lett 3(9):1255–1259
47. Patla I, Acharya S, Zeiri L, Israelachvili J, Efrima S, Golan Y (2007) Synthesis, two-dimensional assembly, and surface pressure-induced coalescence of ultranarrow PbS nanowires. Nano Lett 7(6):1459–1462
48. Deb S (1969) A novel electrophotographic system. App Opt 8(101):192–195
49. Niklasson GA, Granqvist CG (2007) Electrochromics for smart windows: thin films of tungsten oxide and nickel oxide, and devices based on these. J Mater Chem 17(2):127–156
50. Liu BJ-W, Zheng J, Wang J-L, Xu J, Li H-H, Yu S-H (2013) Ultrathin $W_{18}O_{49}$ nanowire assemblies for electrochromic devices. Nano Lett 13(8):3589–3593
51. Nakanishi H, Bishop KJ, Kowalczyk B, Nitzan A, Weiss EA, Tretiakov KV, Apodaca MM, Klajn R, Stoddart JF, Grzybowski BA (2009) Photoconductance and inverse photoconductance in films of functionalized metal nanoparticles. Nature 460(7253):371–375
52. Mueggenburg KE, Lin X-M, Goldsmith RH, Jaeger HM (2007) Elastic membranes of close-packed nanoparticle arrays. Nat Mat 6(9):656–660
53. Maier SA, Kik PG, Atwater HA, Meltzer S, Harel E, Koel BE, Requicha AA (2003) Local detection of electromagnetic energy transport below the diffraction limit in metal nanoparticle plasmon waveguides. Nat Mat 2(4):229–232
54. Nie Z, Petukhova A, Kumacheva E (2010) Properties and emerging applications of self-assembled structures made from inorganic nanoparticles. Nat Nanotech 5(1):15–25
55. Shen X, Chen L, Li D, Zhu L, Wang H, Liu C, Wang Y, Xiong Q, Chen H (2011) Assembly of colloidal nanoparticles directed by the microstructures of polycrystalline ice. ACS Nano 5(10):8426–8433
56. Coe-Sullivan S, Steckel JS, Woo WK, Bawendi MG, Bulović V (2005) Large-area ordered quantum-dot monolayers via phase separation during spin-casting. Adv Mater 15(7): 1117–1124
57. Qian H-S, Yu S-H, Gong J-Y, Luo L-B, Fei L-F (2006) High-quality luminescent tellurium nanowires of several nanometers in diameter and high aspect ratio synthesized by a poly (vinyl pyrrolidone)-assisted hydrothermal process. Langmuir 22(8):3830–3835
58. Ahniyaz A, Sakamoto Y, Bergström L (2007) Magnetic field-induced assembly of oriented superlattices from maghemite nanocubes. Proc Nat Acad Sci 104(45):17570–17574

59. Lee SH, Liddell CM (2009) Anisotropic magnetic colloids: a strategy to form complex structures using nonspherical building blocks. Small 5(17):1957–1962
60. Tang SK, Derda R, Mazzeo AD, Whitesides GM (2011) Reconfigurable self-assembly of mesoscale optical components at a liquid-liquid interface. Adv Mater 23(21):2413–2418
61. Shi HY, Hu B, Yu XC, Zhao RL, Ren XF, Liu SL, Liu JW, Feng M, Xu AW, Yu SH (2010) Ordering of disordered nanowires: spontaneous formation of highly aligned, ultralong Ag nanowire films at oil–water–air interface. Adv Funct Mater 20(6):958–964
62. Korgel BA (2010) Nanocrystal superlattices: Assembly at liquid interfaces. Nat Mat 9:701–703
63. Dong A, Chen J, Vora PM, Kikkawa JM, Murray CB (2010) Binary nanocrystal superlattice membranes self-assembled at the liquid-air interface. Nature 466(7305):474–477
64. Duan H, Wang D, Kurth DG, Möhwald H (2004) Directing self-assembly of nanoparticles at water/oil interfaces. Angew Chem Int Ed 43(42):5639–5642
65. Cheng L, Liu A, Peng S (2010) ACS Nano 4:6098–6104 (2010)
66. Reincke F, Kegel WK, Zhang H, Nolte M, Wang D, Vanmaekelbergh D, Möhwald H (2006) Understanding the self-assembly of charged nanoparticles at the water/oil interface. Phy Chem Chem Phy 8(33):3828–3835
67. Reincke F, Hickey SG, Kegel WK, Vanmaekelbergh D (2004) Spontaneous assembly of a monolayer of charged gold nanocrystals at the water/oil interface. Angew Chem Int Ed 43(4):458–462
68. Ma H, Hao J (2011) Ordered patterns and structures via interfacial self-assembly: superlattices, honeycomb structures and coffee rings. Chem Soc Rev 40(11):5457–5471
69. Wang D, Duan H, Möhwald H (2005) The water/oil interface: the emerging horizon for self-assembly of nanoparticles. Soft Matt 1(6):412–416
70. Liu JW, Zhang SY, Qi H, Wen WC, Yu SH (2012) A general strategy for self-assembly of nanosized building blocks on liquid/liquid interfaces. Small 8(15):2412–2420
71. Qin Y, Wang X, Wang ZL (2008) Microfibre–nanowire hybrid structure for energy scavenging. Nature 451(7180):809–813
72. Sánchez-Iglesias A, Grzelczak M, Pérez-Juste J, Liz-Marzán LM (2010) Binary self-assembly of gold nanowires with nanospheres and nanorods. Angew Chem Int Ed 49(51):9985–9989
73. Ahn J-H, Kim H-S, Lee KJ, Jeon S, Kang SJ, Sun Y, Nuzzo RG, Rogers JA (2006) Heterogeneous three-dimensional electronics by use of printed semiconductor nanomaterials. Science 314(5806):1754–1757
74. Srivastava S, Santos A, Critchley K, Kim K-S, Podsiadlo P, Sun K, Lee J, Xu C, Lilly GD, Glotzer SC (2010) Light-controlled self-assembly of semiconductor nanoparticles into twisted ribbons. Science 327(5971):1355–1359
75. Gates BD, Xu Q, Stewart M, Ryan D, Willson CG, Whitesides GM (2005) New approaches to nanofabrication: molding, printing, and other techniques. Chem Rev 105(4):1171–1196
76. Liu J-W, Xu J, Ni Y, Fan F-J, Zhang C-L, Yu S-H (2012) A family of carbon-based nanocomposite tubular structures created by in situ electron beam irradiation. ACS Nano 6(5): 4500–4507
77. Liu J-W, Wang J-L, Huang W-R, Yu L, Ren X-F, Wen W-C, Yu S-H (2012) Ordering Ag nanowire arrays by a glass capillary: a portable, reusable and durable SERS substrate. Sci Rep 2

Chapter 4
Electron-Beam-Induced Nanowire Assemblies

4.1 Introduction

In material science, self-assembly and nanofabrication have become one of the most active research areas [1–4]. With the help of self-assembly or directed assembly techniques, numerous kinds of nanostructures have been successfully obtained, which provides a way to achieve a unique combination of desired electronic, magnetic, and optical properties that are often not available in single-component materials [5–8]. By using various stimuli, self-assembly process can be initiated, including external forces [9–12], phase interface [13], evaporation of the solvent [14, 15], self-assembly directed by small molecules [16, 17], and changing the color and intensity of incident light [18, 19], electron-beam irradiation [20], photochemical triggering [21, 22]. The irradiation conditions can develop a situation where the solid is far from thermal equilibrium, and under certain circumstances, this situation leads to direct interactions between incident electrons and materials, which results in the formation of self-assembled structures [23, 24]. Furthermore, the transmission electron microscopy that is most frequently used for all kinds of nanostructure investigation also has the advantage to cause a structural variation of the specimen by electron irradiation and lead to the fabrication of unpredicted and very exciting structures [25, 26]. Despite the massive success so far, assembly of nanosized building blocks and nanofabrication is still challenging and shows a certain deficiency for a sufficient level of control; hence, more efforts are required for further development [27]. Recently, Yu and coworkers transformed the conductive carbon film with a thickness of 10 nm into a tubular structure by electron-beam irradiation, thereby directly assembling the nanomaterials such as nanoparticles, nanorods, nanowires, and monolayer nanoplates on the conductive carbon film [28]. There is a wide range of materials that can be designed to produce carbon-based tubular structures and can be suitable candidates for functional materials as well as related applications. The effects of electron irradiation on carbon nanostructures have been studied previously [29–31]. Ugarte et al. (1992) described the crimping and closing

J. Liu, *Well-Organized Inorganic Nanowire Films*, Springer Theses,
DOI 10.1007/978-981-10-3947-8_4

of graphite networks under electron-beam irradiation. Under TEM electron-beam irradiation, the top temperature of the carbon film is different from the bottom end of the carbon film, which offers significant strain, resulting in bending the nanofilm into a cylindrical shape. To study the properties of this process, single Ag nanowires, low concentration disordered Ag nanowires, and ordered Ag nanowires assembled by LB techniques are selected as the elastic material, and mechanical energy was stored by structural memory to achieve potential application [32].

4.2 Fabrication of Carbon Films by In Situ Electron-Beam Irradiation

To demonstrate the versatility of this process, four different dimensions of nanocrystals, namely nanorods, nanoparticles, nanosheets, and nanowires, are chosen. The schematic design of this process is presented in Fig. 4.1, showing that TEM copper mesh (300 mesh) is provided with a 10-nm-thick carbon film on the surface without the need for further processing or functionalization. On exposure to electron-beam irradiation after cutting a thin slit on the sample, it begins to bend into an approximately cylindrical surface and then gradually into a tubular structure due to the strain relief. In addition, if the carbon film supported-side of TEM grid is loaded with nanomaterials such as nanoparticles, nanorods, nanowires, and

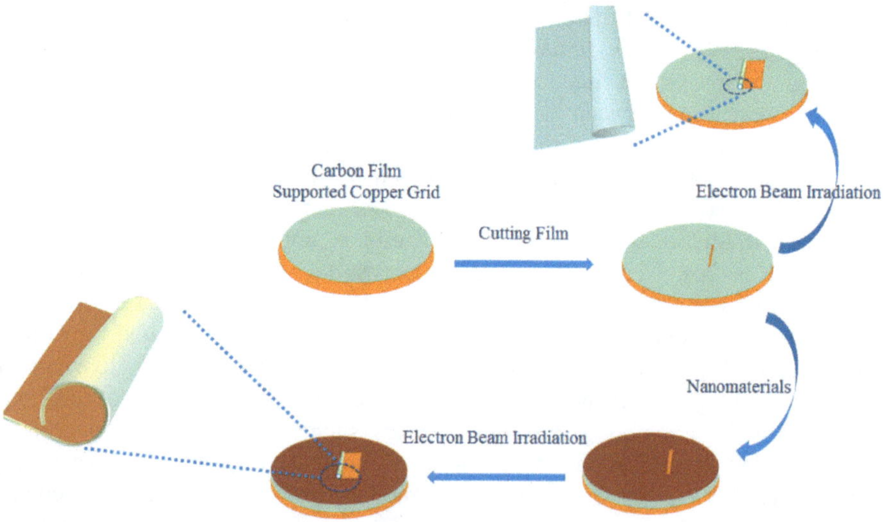

Fig. 4.1 Schematic design of carbon-based solid films rolled up into nanotubes under the electron-beam irradiation. Reprinted with the permission from Ref. [28]. Copyright 2012 American Chemical Society

Fig. 4.2 **a–d** Curling of carbon film with time going on under the electron-beam irradiation of TEM at 100 kV at the time interval is 1 s. **e, f** TEM images of the cruled carbon film tubes formed with the membrane slites of differencet size and shape. Reprinted with permission from Ref. [28]. Copyright 2012 American Chemical Society

monolayer nanoplatelets, they will bend together with the carbon film under the irradiation of the electron beam. The bottom end of the carbon film is believed to provide critical strain to force the nanomaterial film to bend into a cylindrical shape.

The relationship between the carbon films bend image and the time is presented in Fig. 4.2a–d, showing that when the time is extended at intervals of 1 s at an accelerating voltage of 100 kV [28]. If the electron-beam irradiation time is long enough, a multilayer tubular structure will be produced as shown in Fig. 4.2e, f. When the nanomaterials are added on the surface of the carbon film, the bending speed of the carbon film will slow down. The shape and size of carbon membrane slit are the most important prompting factor in controlling the diameter of the tubular structure (Fig. 4.2). In addition to this factor, there are many other factors that can influence the diameter of the tubular structure such as current density, operating voltage, and the introduction of additional nanomaterials [28].

4.3 Fabrication of Carbon-Based Nanocomposites by In Situ Electron-Beam Irradiation

On loading the Te nanowires on carbon films, curling of carbon film occurs with time prolonging at 100 kV as shown in Fig. 4.3a–d [28]. When the electron beam is focused, the solid film begins to curl in the detaching area of the substrate, indicating that this process can be well adapted to some natural systems. A similar process of transformation is observed in organisms, where the interaction between the membrane molecules changes the lateral tension of the lipid duplex. Figure 4.3a

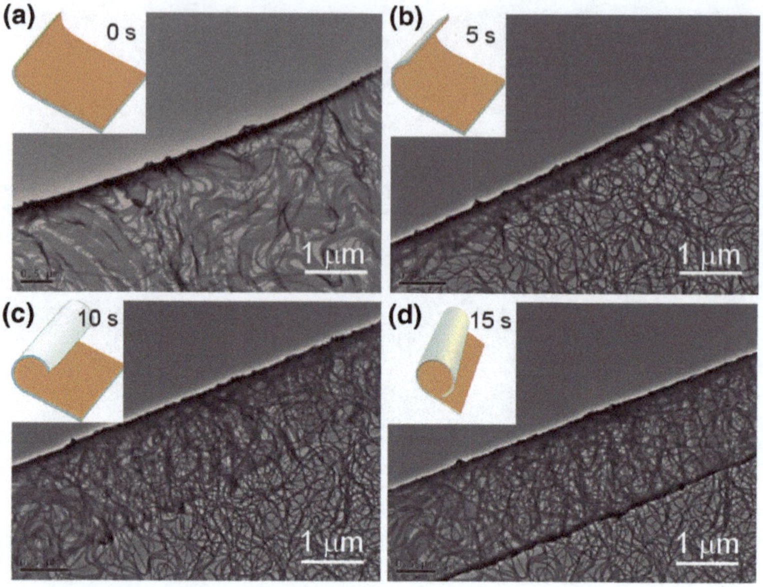

Fig. 4.3 a–d Time-dependent curling assembly of a carbon film with nanowires under the irradiation of a TEM beam at 100 kV. Reprinted with permission from Ref. [28]. Copyright 2012 American Chemical Society

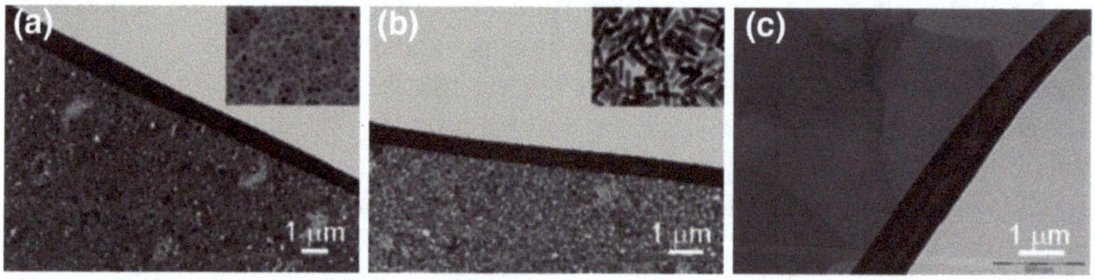

Fig. 4.4 a–c Carbon film curling with monodisperse Cu_2ZnSnS_4 nanoparticles, Au nanorods, and graphene. Reprinted with permission from Ref. [28]. Copyright 2012 American Chemical Society

shows that the most pronounced bending, especially the tubular structure, begins to form after the 5 s that changes to nanowire-based tube-like structure on further exposure to irradiations. Like other emerging research areas, researchers working in the nanowire assembly field are also trying to develop new complex structures to expand the desired applications. As a general and versatile assembly method, this technique can be extended to criminate many other kinds of nanomaterials with different dimensionalities such as nanoparticles, nanorods, and nanowires [28].

Similarly, by using the same approach zero-dimensional nanoparticles with a diameter of 10 nm, and two-dimensional nanosheets (e.g., graphene), and nanorods with a diameter of width 19 nm and a length of 78 nm, can be assembled (Fig. 4.4a–c) [28]. The transition to kinetic energy and elastic potential energy is of critical importance in the process of device processing to novel new functions. In order to store the mechanical energy into a film for potential applications, silver nanowires are introduced as the elastic metal objects with structural memory. Three types of films are prepared by using single silver nanowire, disordered Ag nanowires with low density, and an ordered Ag-nanowire film assembled with LB are introduced to the carbon film supported-side of TEM grid with a slit, in order to investigate the power of the in situ rolling process. In case of single nanowire film with a diameter of 60 nm and a length of 15 μm, a rapid curling of film is observed at the time interval of 5 s, whereas, in case of disordered nanowires, the rolling process becomes more difficult and the time taken becomes 10 s Fig. 4.5d–f, indicating that the rolling process becomes more complicated as the number of nanowires increases. Ordered Ag nanowires obtained by the LB technique can be curled with carbon film from the lateral direction, but they are unaffected from the axial direction (Fig. 4.5g) [28].

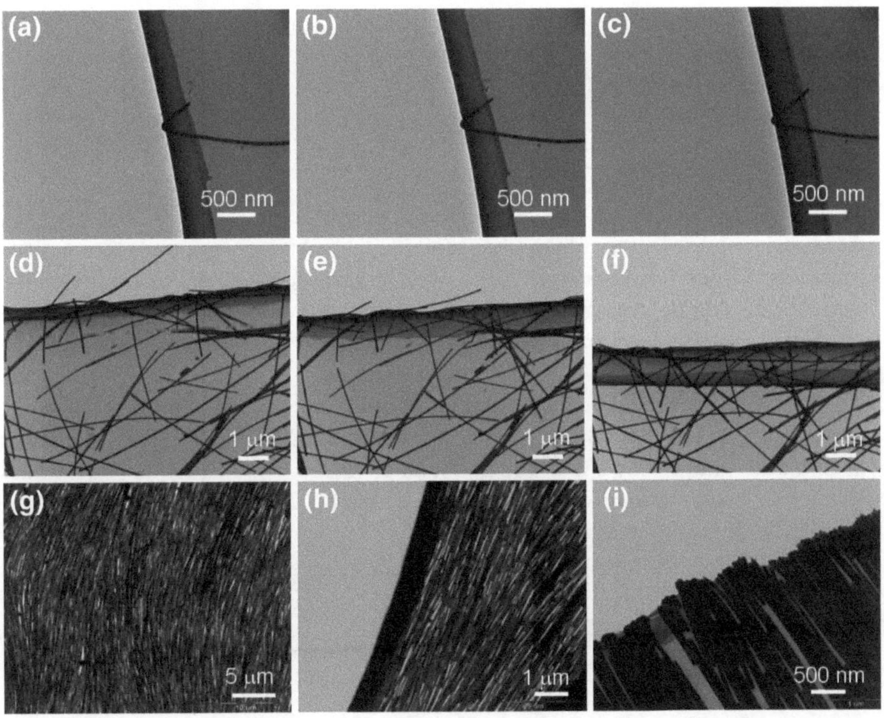

Fig. 4.5 **a–c** TEM images of time-dependent curling of a carbon film with a single Ag nanowire for a time interval of 5 s. **d–f** Curling of a carbon film with disordered Ag nanowires with lower density at the time interval of 10 s. **g** Ordered Ag nanowires assembled by LB with higher density. **h** TEM images of carbon film can curl with ordered Ag nanowires from the lateral direction, but they cannot curl from axial direction (**i**). Reprinted with permission from Ref. [28]. Copyright 2012 American Chemical Society

In order to utilize this phenomenon of tubular building blocks, it is essential to understand the mechanism of rolling process [33, 34]. A system away from equilibrium can reveal complex transitory structures with curly structures, and once unconstrained from their substrate, the layer rolls up by itself and forms a nanotube [35–39]. It is clear from above discussion that during film formation on the surface of the film or during the curling of the film, mechanical deformation occurs due to the strain difference between the different layers in the film system. When two mismatched epitaxial thin-film layers are forced to be on the same substrate, a large residual strain energy is stored in the bilayer film. Moreover, the film curls to form various three-dimensional structures on releasing from the substrate to reduce stored strain energy, including nanotubes and microtubes.

With the assumption that the optimal release of the elastic energy stored in the carbon film under the irradiation of the electron beam is the major factor for the formation of hybrid tabular structure, the curvature of the curved tubular structure can be estimated by a multilayer hinge construction model containing residual strain [40]. For bilayers, there are bottom and top layers with thicknesses $h1$ and $h2$, respectively, and $\varepsilon 1$ and $\varepsilon 2$ are the biaxial residual strains of the two films, respectively. The separation area of the substrate is denoted by $W \times L$, and W and L are the length and width of the tear rip, respectively. The radius of the tubular structure can be written as follows:

$$R = \frac{h_1\left[1+\beta^2\alpha^4+2\beta\alpha\left(2+2\alpha^2+3\alpha\right)\right]}{6\eta\beta\alpha\left(1+\alpha\right)\left(\varepsilon_2-\varepsilon_1\right)} \tag{4.1}$$

Here, L is a form factor. When L is small, $\eta = 1$, which corresponds to the plane compression; when L is large, $\eta = 1 + \nu$, which corresponds to the plane stretching. The value of η can be determined by finite factor simulation between the two cases. From Eq. 4.1, it is clear that when $\varepsilon 2 - \varepsilon 1 > 0$, film is in the bending state. In addition, on regulating the value of α and β, the radius of the tube can be adjusted through the mixing process. It is important to note that when the bilayers are in the compressed state ($\varepsilon 2 < 0$, $\varepsilon 1 < 0$), it will tend to produce wrinkles. The calculation [41] in the literature shows that larger W and L values and smaller $\varepsilon 2 - \varepsilon 1$ values will tend to produce wrinkles. Therefore, a careful control is required for the separation process, to prevent the formation of wrinkles to generate the tubular structure.

4.4 Conclusions

In summary, a unique technique is discussed to directly and specifically roll up the carbon films into tubular structures, loaded with different nanostructures such as nanoparticles, nanowires, nanorods, and nanosheets by taking the advantage of electron-beam-induced surface stress. This strategy has proven to be a versatile way by which various nanomaterials of different shapes and sizes can be co-assembled with a conductive carbon film and can be bent into a tubular structure. This excellent design flexibility approach is expected to produce different tubular structures and composites with encouraging applications, including encapsulation, nanoscale fluid delivery, and capillarity on the nanometer scale as this technique has the ability to load various nanomaterials with different shape and size.

References

1. Scholes GD, Rumbles G (2006) Excitons in nanoscale systems. Nat Mat 5(9):683–696
2. Michalet X, Pinaud F, Bentolila L, Tsay J, Doose S, Li J, Sundaresan G, Wu A, Gambhir S, Weiss S (2005) Quantum dots for live cells, in vivo imaging, and diagnostics. Science 307 (5709):538–544
3. Daniel M-C, Astruc D (2004) Gold nanoparticles: assembly, supramolecular chemistry, quantum-size-related properties, and applications toward biology, catalysis, and nanotechnology. Chem Rev 104(1):293–346
4. Gates BD, Xu Q, Stewart M, Ryan D, Willson CG, Whitesides GM (2005) New approaches to nanofabrication: molding, printing, and other techniques. Chem Rev 105(4):1171–1196
5. Overgaag K, Evers W, de Nijs B, Koole R, Meeldijk J, Vanmaekelbergh D (2008) Binary superlattices of PbSe and CdSe nanocrystals. J Am Chem Soc 130(25):7833–7835
6. Urban JJ, Talapin DV, Shevchenko EV, Kagan CR, Murray CB (2007) Synergism in binary nanocrystal superlattices leads to enhanced p-type conductivity in self-assembled PbTe/Ag2Te thin films. Nat Mat 6(2):115–121
7. Shevchenko EV, Talapin DV, Kotov NA, O'Brien S, Murray CB (2006) Structural diversity in binary nanoparticle superlattices. Nature 439(7072):55–59
8. Kalsin AM, Fialkowski M, Paszewski M, Smoukov SK, Bishop KJ, Grzybowski BA (2006) Electrostatic self-assembly of binary nanoparticle crystals with a diamond-like lattice. Science 312(5772):420–424
9. Yao HB, Fang HY, Tan ZH, Wu LH, Yu SH (2010) Biologically inspired, strong, transparent, and functional layered organic–inorganic hybrid films. Angew Chem Int Ed 122(12):2186–2191
10. Liu J-W, Zhu J-H, Zhang C-L, Liang H-W, Yu S-H (2010) Mesostructured assemblies of ultrathin superlong tellurium nanowires and their photoconductivity. J Am Chem Soc 132 (26):8945–8952
11. Srivastava S, Kotov NA (2008) Composite layer-by-layer (LBL) assembly with inorganic nanoparticles and nanowires. Acc Chem Res 41(12):1831–1841
12. Hu M-J, Lu Y, Zhang S, Guo S-R, Lin B, Zhang M, Yu S-H (2008) High yield synthesis of bracelet-like hydrophilic Ni–Co magnetic alloy flux-closure nanorings. J Am Chem Soc 130 (35):11606–11607
13. Shi HY, Hu B, Yu XC, Zhao RL, Ren XF, Liu SL, Liu JW, Feng M, Xu AW, Yu SH (2010) Ordering of disordered nanowires: spontaneous formation of highly aligned, ultralong Ag nanowire films at oil–water–air interface. Adv Funct Mater 20(6):958–964
14. Maheshwari S, Zhang L, Zhu Y, Chang H-C (2008) Coupling between precipitation and contact-line dynamics: multiring stains and stick-slip motion. Phy Rev Lett 100(4):044503
15. Ming T, Kou X, Chen H, Wang T, Tam HL, Cheah KW, Chen JY, Wang J (2008) Ordered gold nanostructure assemblies formed by droplet evaporation. Angew Chem Int Ed 120 (50):9831–9836

16. Song G, Chen C, Qu X, Miyoshi D, Ren J, Sugimoto N (2008) Small-molecule-directed assembly: a gold nanoparticle-based strategy for screening of homo-adenine DNA duplex binders. Adv Mater 20(4):706–710
17. Zhao Y, Thorkelsson K, Mastroianni AJ, Schilling T, Luther JM, Rancatore BJ, Matsunaga K, Jinnai H, Wu Y, Poulsen D (2009) Small-molecule-directed nanoparticle assembly towards stimuli-responsive nanocomposites. Nat Mat 8(12):979–985
18. Fava D, Winnik MA, Kumacheva E (2009) Photothermally-triggered self-assembly of gold nanorods. Chem Comm 18:2571–2573
19. Zhou J, Sedev R, Beattie D, Ralston J (2008) Light-induced aggregation of colloidal gold nanoparticles capped by thymine derivatives. Langmuir 24(9):4506–4511
20. Qi X, Xue C, Huang X, Huang Y, Zhou X, Li H, Liu D, Boey F, Yan Q, Huang W (2010) polyphenylene dendrimer-templated in situ construction of inorganic-organic hybrid rice-shaped architectures. Adv Funct Mater 20(1):43–49
21. Kim WJ, Kim SJ, Lee K-S, Samoc M, Cartwright AN, Prasad PN (2008) Robust microstructures using UV photopatternable semiconductor nanocrystals. Nano Lett 8 (10):3262–3265
22. Smith AR, Watson DF (2009) Photochemically triggered assembly of composite nanomaterials through the photodimerization of adsorbed anthracene derivatives. Chem Mat 22 (2):294–304
23. Son JG, Chang J-B, Berggren KK, Ross CA (2011) Assembly of sub-10-nm block copolymer patterns with mixed morphology and period using electron irradiation and solvent annealing. Nano Lett 11(11):5079–5084
24. Warner JH (2008) Self-assembly of ligand-free PbS nanocrystals into nanorods and their nanosculpturing by electron-beam irradiation. Adv Mater 20(4):784–787
25. de Jonge N, Ross FM (2011) Electron microscopy of specimens in liquid. Nat Nanotech 6 (11):695–704
26. Zheng H, Smith RK, Jun Y-W, Kisielowski C, Dahmen U, Alivisatos AP (2009) Observation of single colloidal platinum nanocrystal growth trajectories. Science 324(5932):1309–1312
27. Dong Z, Lai X, Halpert JE, Yang N, Yi L, Zhai J, Wang D, Tang Z, Jiang L (2012) Accurate control of multishelled ZnO hollow microspheres for dye-sensitized solar cells with high efficiency. Adv Mater 24(8):1046–1049
28. Liu J-W, Xu J, Ni Y, Fan F-J, Zhang C-L, Yu S-H (2012) A family of carbon-based nanocomposite tubular structures created by in situ electron beam irradiation. ACS Nano 6 (5):4500–4507
29. Mei Y, Thurmer DJ, Deneke C, Kiravittaya S, Chen Y-F, Dadgar A, Bertram F, Bastek B, Krost A, Christen JR (2009) Fabrication, self-assembly, and properties of ultrathin AlN/GaN porous crystalline nanomembranes: tubes, spirals, and curved sheets. ACS Nano 3(7):1663–1668
30. Banhart F (2006) Irradiation of carbon nanotubes with a focused electron beam in the electron microscope. J Mat Sci 41(14):4505–4511
31. Banhart F (1999) Irradiation effects in carbon nanostructures. Rep Prog Phy 62(8):1181
32. Ugarte D (1992) Curling and closure of graphitic networks under electron-beam irradiation. Nature 359:707–709
33. Gunawidjaja R, Peleshanko S, Ko H, Tsukruk VV (2008) Bimetallic nanocobs: decorating silver nanowires with gold nanoparticles. Adv Mater 20(8):1544–1549
34. Chang S, Ko H, Gunawidjaja R, Tsukruk VV (2011) Raman markers from silver nanowire crossbars. J Phy Chem C 115(11):4387–4394
35. Schmidt OG, Eberl K (2001) Nanotechnology: thin solid films roll up into nanotubes. Nature 410(6825):168–168
36. Shenoy VB, Reddy CD, Zhang Y-W (2010) Spontaneous curling of graphene sheets with reconstructed edges. ACS Nano 4(8):4840–4844
37. Schmidt O, Schmarje N, Deneke C, Müller C, Jin-Phillipp NY (2001) Three-dimensional nano-objects evolving from a two-dimensional layer technology. Adv Mater 13(10):756–759

38. Cho JH, James T, Gracias DH (2010) Curving nanostructures using extrinsic stress. Adv Mater 22(21):2320–2324
39. Guo X-H, Yu S-H, Lu Y, Yuan G-B, Sedlák M, Cölfen H (2010) Spontaneous formation of hierarchically structured curly films of nickel carbonate hydrate through drying. Langmuir 26 (12):10102–10110
40. Nikishkov GP (2003) Curvature estimation for multilayer hinged structures with initial strains. J App Phy 94(8):5333–5336
41. Banhart F (2006) Irradiation of carbon nanotubes with a focused electron beam in the electron microscope. J Mater Sci 41(14):4505–4511

Chapter 5
Applications of the Nanowire Assemblies

5.1 Introduction

When the diameter of a nanowire structure is close to or less than the length of a solid state, such as the Bauer radius, the optical wavelength, the phonon mean free path, the critical size of the magnetic field, and the diffusion length of the exciton, many physical properties of the nanowire have an important relationship with the surface [1]. For the direct application of the nanowire system, it is important to study the ordered and interconnected functional nanowires with complex and multiscale structures [2]. In the following sections, we will describe a series of applications of nanowire structures, such as electrical, optical, and photoconductive applications. With the utilization of well-arranged nanowire structure, numerous functional devices and nanoelectrical devices can be fabricated [3–9]. Moreover, the nanowire position and the interconnections within the nanowires are very important in order to create special properties, which develop their importance in the manufacturing of various devices. In recent years, many complex and multiscale nanowire structures and their applications have been studied. More recently, the design, fabrication, and the use of a programmable nanoprocessor that can overcome these challenges have been described.

Duan et al. have fabricated high-speed graphene transistors to achieve this superperformance by self-assembling nanowires [10]. First, by using graphene, this physical assembly provides large mobility. In addition, this self-assembly process ensures precise placement of the edge of the gate electrode, so there is no significant overlap or gap between the electrodes, minimizing the resistance. Myoung et al. have published a process for the fabrication of field-effective Si nanowire transistors with programmable nanowires that have homogeneous transport properties. Depending on the number of nanowires in the channel width of the transistor, the nanowire bridges between the electrodes can be divided into three groups of PDMS blocks [10]. First, ordered Si nanowires can be evenly coated on a flexible plastic substrate. The FET is then fabricated from the transformed nanowires, through

J. Liu, *Well-Organized Inorganic Nanowire Films*, Springer Theses,
DOI 10.1007/978-981-10-3947-8_5

microprocessor technology and low-power operation. According to the assembly of nanowires, the sensor array of flexible plastic substrates shows a high degree of sensitivity.

5.2 Fabrication of Nanowire Stamper

Ordered 2D functional nanostructure arrays have a wide range of applications [11]. With the use of Te nanowires, nanochannels can be fabricated to a poly(methyl methacrylate) (PMMA) substrate by following a simple pattern transfer method [12]. The schematic design for the fabrication of mesostructured nanopattern by creating a monolayer of Te nanowires is shown in Fig. 5.1a. Initially, the glass slides are ultrasonically washed in acetone for 10 min and then purge with ultrapure water. On the surface of the glass slide, Te monolayer is formed, which is used as a stamp for nanocontact printing. Afterward, another slide is coated with PMMA by a spin coating that further put onto a Te nanowire monolayer stamp by applying suitable pressure. After heating at 100 °C for 1 h, the PMMA-coated glass slide substrate is rapidly pulled away from the stamp, resulting nanochannels of Te nanowires on the surface of poly(methyl methacrylate) (PMMA) substrate [12].

During the transfer process, no obvious change is observed in the patterns of Te nanowire monolayer which is demonstrated by XRD experiments (Fig. 5.1b, c). This lightens the stability of the mesostructures of the Te nanowire monolayer during the transfer process. Furthermore, the SAXRD pattern of PMMA film illustrates a change in the width of PMMA channel from 6.3 to 6.7 nm, after being

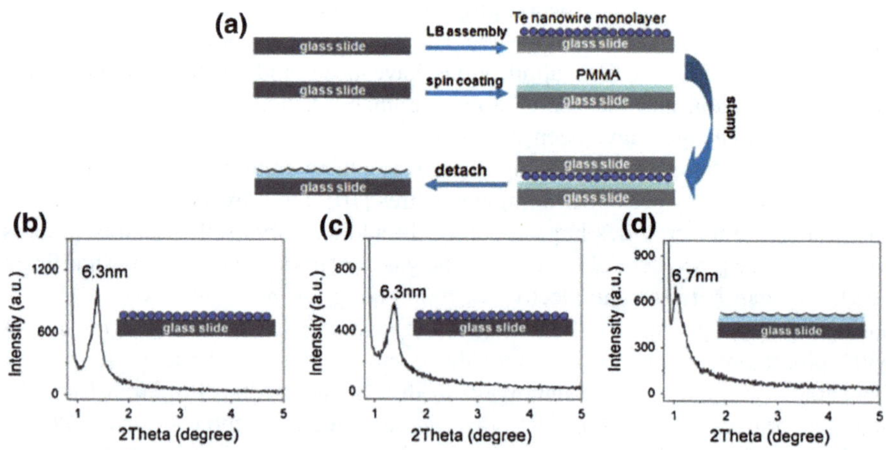

Fig. 5.1 Schematic design for fabricating **a** Te monolayer stamp for nanopatterns. **b**, **c** Te nanowires before monolayer and after the imprinting. **d** The PMMA substrate with constant nanopatterns. Reprinted with the permission from Ref. [12]. Copyright 2010 WILEY-VCH

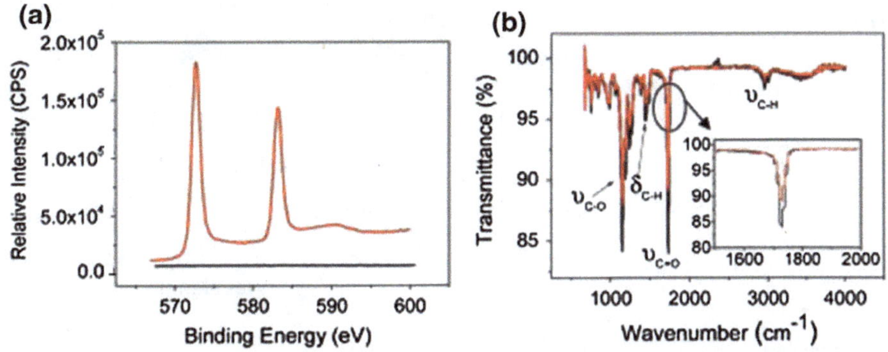

Fig. 5.2 **a** Te 3D spectrum of the PMMA film, **b** FTIR spectra of the PMMA film. Reprinted with the permission from Ref. [12]. Copyright 2010 WILEY-VCH

detached from another slide (Fig. 5.1d). The other techniques such as XPS and IR also confirm the successful transfer process (Fig. 5.2a, b) [13].

5.3 Nanowire Selective Transmission

The transmission of polarized light through periodic nanostructures has attracted huge attention because of exciting applications [14, 15]. The incident light polarization determines the excitation mode of surface plasmon, associated with the periodic structure. Figure 5.3 shows the schematic design of experimental setup with a plot, indicating the transmission efficiencies of different wavelength light for different samples Fig. 5.3b [16]. The transmission efficiency mainly depends on two factors that are nanowire order and incident light polarization, attributed to the surface plasmon. With the help of optical fiber spectrometer, the transmission spectra of the glass capillaries with ordered silver nanowires on the inner walls are recorded as shown in Fig. 5.3b.

In a particular arrangement, the white light is passed through a single-mode fiber and a polarizer (only vertically polarized light can pass) from a stabilized tungsten–halogen source that illuminates the sample. The polarization angle θ presents the angle between the incident light and the longitudinal axis of the capillary. In order to incident light on the sample, the sample is arranged between two lenses of 35 mm focal length. The light emitting from the sample is set to move into the spectrograph with the anticlockwise rotation (with the rotation angle from 0° to 360°) of the sample in the plane perpendicular to the illuminating light. Figure 5.3c shows the transmission spectra of the sample with the rotation angle of 0° and 90°. 18 spectra were recorded from angle 0° to 360°, and the peak intensity of integral of the spectra is found at 650 nm (Fig. 5.3c), indicating the light intensity of the transmitted spectrum as a function of the rotation angle. From above data, it is

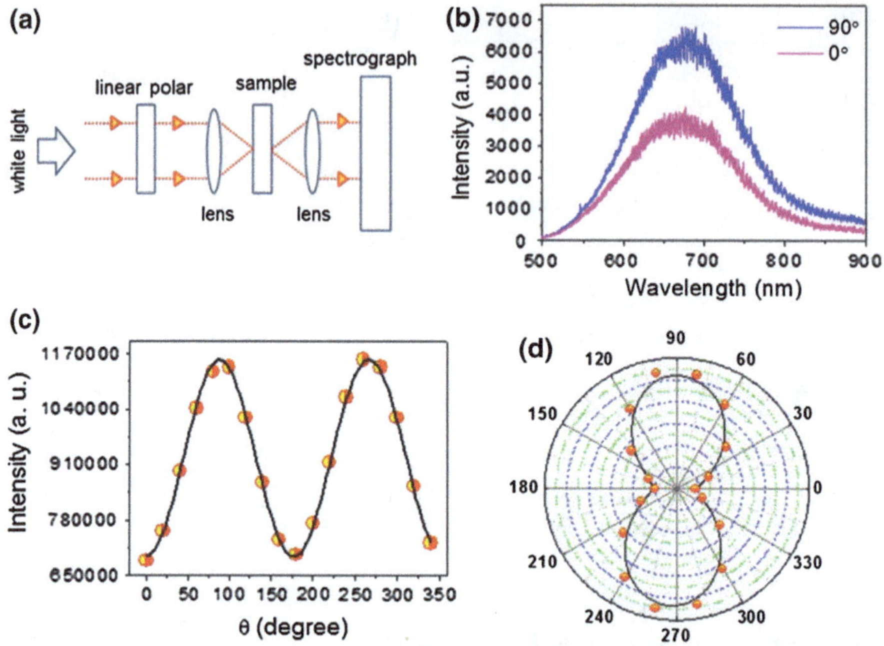

Fig. 5.3 **a** Schematic illustration of the experimental setup. **b** Transmission spectra of the film of silver nanowires with the rotation angles of 0° and 90°. **c** Intensity of the transmittance as a function of rotation angles h. **d** Transformation of the data in **c** into polar coordinates. Reprinted with the permission from Ref. [16]. Copyright 2012 Nature Publishing Group

concluded that stimulation of surface plasmon polaritons (SPPs) occurs only when the light is perpendicular to nanowire alignment, making the vertical component of the electric field in working state that is $E \sin \theta$. However, the energy relationship should be $E^2 \sin^2 \theta$. While considering the background factor, the function of mutual relations should be as follow.

$$S = a\,(\sin^2\theta + b)$$

5.4 Nanowire Photoconductors

Photoconductivity is an optical and electrical phenomenon that causes the material to have a greater electrical conductivity due to the absorption of electromagnetic radiation, such as visible light, ultraviolet light, infrared light, or radioactive rays. Recently, nanostructure materials have attracted more attentions because of their optical properties such as surface effects, holes, geometrical confinement of photons, and quantum confinement of electrons. In semiconducting materials, element selenium was explored as the first photoconductive element, and then, the further investigation was carried out on other photoconductive materials gradually. Furthermore, semiconducting tellurium having a bandgap of 0.3 eV is considered as an important narrow-bandgap semiconductor having outstanding applications in future highly efficient photoconductor devices. The photoconductivity effect was initially reported by Bartlett [17], before that the photoconductivity of Te was considered unaffected by radiation. Recently, the photoconductive properties of aligned Te nanowires have been studied by Yu and coworkers [18]. By using the Lb technique, different layers of Te nanowires with the same axial direction are initially fabricated on a Si/SiO_2 substrate [19, 20] with the thermal evaporation of Au electrode through positioning a micrometer-sized Au wire on the nanowires as the mask to attain a gap between two electrodes. The current–voltage curves measured in dark and white light brilliance are shown in Fig. 5.4b, showing linear behavior that exhibits the ohmic contact feature. Moreover, on turning the light on from the dark, the electrical resistance of the nanowire film decreases, similar to other photoconduction material. It is evident that the current of the devices increases considerably with the increase in light intensity and the current increases linearly with increasing illumination power (Fig. 5.4b inset). The increase in conductivity with light allows to reversibly switch the nanowire films between "off" and "on" states, but the performance is highly dependent on the thickness of the nanowire films, from monolayer, three layers, five layers, and 20 layers of Te nanowire films (Fig. 5.4c–f).

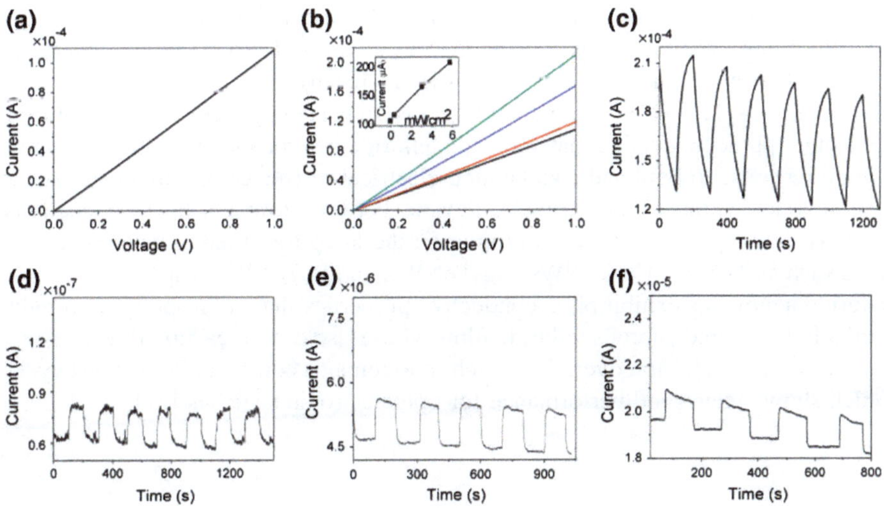

Fig. 5.4 **a** Schematic illustration of the device based on Te sample for the photoconductive test. **b** *I*–*V* curves measured under different light intensities. **c** Reversible switching of a 20-layer Te nanowire device. **d–f** Reversible switching photoconductive property of the Te nanowire films with one, three, and five layers, respectively. Reprinted with the permission from Ref. [18]. Copyright 2010 American Chemical Society

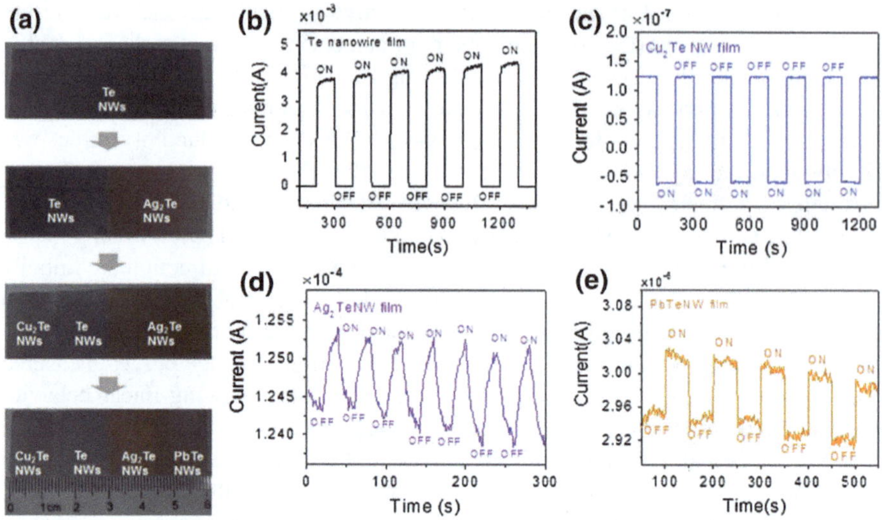

Fig. 5.5 **a** Images of chemical conversion of a TeNW monolayer into Te–Ag_2TeNW films, Cu_2Te–Te–Ag_2TeNW films, and Cu_2Te–Te–Ag_2Te–PbTeNW films. **b–e** Reversible switching of different telluride nanowire patterns between low and high conductivity states. Reprinted with the permission from Ref. [1]. Copyright 2012 WILEY-VCH

In order to expand the photoconductive study of nanowire films, other nanowire films such as Cu_2TeNW, Ag_2TeNW, and PbTeNW were fabricated [21, 22]. Figure 5.5 shows images of the conversion of Te nanowire film into other metal telluride heteronanowire films such as the formation of Te–Ag_2Te heteronanowire films from TeNWs and then Cu_2Te–Te–Ag_2Te heteronanowire films, and finally, Cu_2Te–Te–Ag_2Te–PbTe heteronanowire films, indicated by a color change of film. Like other photoconductive materials, the telluride heteronanowire films also show a good response to light and change their electrical resistance on illumination. The reversible switching of the nanowire devices between low and high conductivity states is shown in Fig. 5.5b–e on turning on the lamp for different heteronanowire films such as TeNWs, Cu_2TeNWs, Ag_2TeNWs, and PbTeNWs, respectively. These telluride nanowires exhibit photoconductive properties that are responsive to light, particularly ordered cuprous telluride films whose rise or fall reaction time does not exceed one second. Moreover, the switch ratio remains constant after the test cycle, which shows their good performance for optoelectronic switches [21].

5.5 Nanowire Memory Device

The most important characteristic of the memory resistance is the dependence of the direction of the electric field; i.e., the resistance of the system can be changed with the voltage or current applied by the system to form a current–voltage loop [23, 24]. For the fabrication of a promising memory device based on resistive switching behavior, a 15-layer ordered silver telluride nanowire film has been prepared by Yu group on a silicon/silicon dioxide (silicon dioxide 500 nm) substrate by the pattern transfer method with Ti- and Au-plated electrode [21]. Figure 5.6a shows the current–voltage curve of nanowire electrodes measured at room temperature for 10 consecutive times. The applied bias voltage gradually changes from the negative maximum voltage to the positive maximum voltage and then back to the negative maximum voltage. Figure 5.6b shows the current–voltage curve in the logarithm of the current. The symmetry and nonlinearity of curve show that the current at the probe voltage for the high-conducting state is greater than the current in the low-conducting state at the same voltage. Figure 5.6c demonstrates the existence of high and low resistance under positive and negative voltage phenomenon as pulses

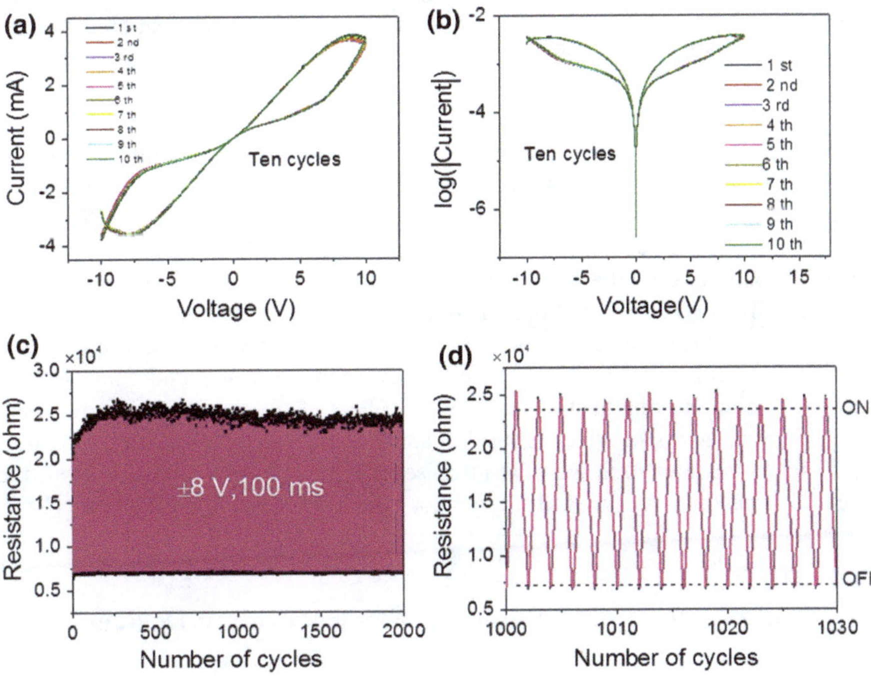

Fig. 5.6 **a** Current–voltage characteristics of 15 layers of well-patterned Ag_2TeNW. **b** The I–V curves with the current in a logarithmic scale. **c** Read-only memory applications over about 2000 cycles. **d** Read-only memory applications from 1000 to 1030. Reprinted with the permission from Ref. [1]. Copyright 2012 WILEY-VCH

of +8 and −8 V (0.1 s) lead to different resistance states, respectively, and a voltage of 5 mV is used as the read voltage.

Moreover, the nanowire memory device shows no state disturbance by continuously applying electric pulses for 2000 cycles, indicating that no overlap between the high-resistance state (HRS) and the low-resistance state (LRS) is found even at 2000 cycles. In 1971, Chua proposed the concept of memory resistance by analyzing the mathematical relationship among the basic variables of several groups of circuits. Eventually, the dynamic properties of electrons and ions are the basis of memory effects. Memory effect is widespread in nature and especially in the nanometer scale, where many impressive and comprehensive reviews related to this topic have been published [18, 25–27]. The reversible current–voltage behavior of this Au/Ag_2Te/Au system can be described by the movement of the electron hole in the electric field, which explains from a low-resistance state to a high-resistance state from a positive voltage. Negative voltages accumulate holes on the surface, resulting in a reduction in the height and width of the Schottky boundary [28].

To change the composition and properties of the materials, cation exchange reactions are widely used by changing the position of cations within the different nanocrystal lattice with a different metal ion [29–31]. Figure 5.7a shows a schematic representation of an ordered nanowire film that changes composition by ion exchange reaction [21]. The compositional control of a nanowire monolayer oriented by chemical transformation is shown in Fig. 5.7b, showing that the films are glossy, flexible, and very smooth. Furthermore, this pattern transfer process is appeared to be a versatile method, and brown Ag_2TeNW patterns can be easily drawn on the dark blue TeNW film with the use of aqueous $AgNO_3$ solution as ink Fig. 5.7c. In order to understand the degree of compositional control, various electrical measurements are conducted on the same nanowire films during a series of sequential chemical transformation pattern transfers (Fig. 5.7d–g). The I–V curves of the sample show that the surface area of Ag_2Te to Te varies from 0 to 8% and 100% exhibiting a different memory effect for both Te–Ag_2Te heteronanowire films and pure Ag_2Te (Fig. 5.7d–g). The boundary of tellurium and silver telluride nanowire heterostructures can be moved by the electron beam of the transmission electron microscope, which is the reason for silver ion movement under the electron beam irradiation. In order to study the memory effect of silver telluride nanowires in detail, different layers of silver telluride nanowire films are prepared as shown in (Fig. 5.7h–k), showing that with the increase in silver telluride nanowire layer, the current also becomes larger, and the memory curve becomes more obvious [21].

5.6 Ordered Nanowire Films for Electrochromic Devices

The electrochromic phenomenon is a reversible process [25–29] caused by the color change due to the adsorption and desorption of some specific particles. This electrochemical process is related to the electron absorption band in the visible region, but also to the redox state. As a well-known inorganic electrochromic material,

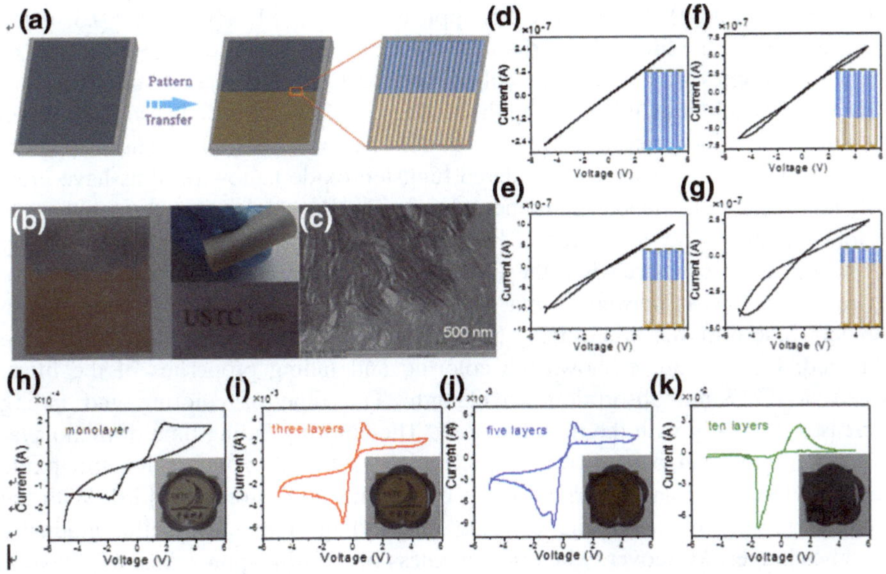

Fig. 5.7 **a** Explanation of compositional control by the chemical transformation of an oriented nanowire monolayer. **b** Pattern transfer monolayer on a flexible substrate. **c** TEM images of a Te–Ag_2Te heteronanowire monolayer. **d–g** A series of sequential measurements on the same nanowire films at the surface areas of **d** 0%, **e** 40%, **f** 60%, and **g** 80%. **h–k** Current–voltage features from −5 to +5 V and from +5 to −5 V of films composed of **h** monolayer, **i** three layers, **j** five layers, and **k** ten layers. Reprinted with the permission from Ref. [21]. Copyright 2012 WILEY-VCH

tungsten oxide exhibits a colorless and blue alternating change with the applied positive and negative voltage changes. This phenomenon has been extensively studied since Deb's discovery in 1969 [30–32]. Unlike single nanowires or disordered arrays, ordered nanofilms have a much larger face ratio and can be used in practical applications to produce sensors that are more controllable, sensitive, and stable. The electrochromic performance of the ordered nanowire films is highly reliant on the thickness of the nanowire films. The schematic design of reversible electrochromic $W_{18}O_{49}$ nanowire film device is shown in Fig. 5.8a, having ordered $W_{18}O_{49}$ nanowire films on ITO substrate without any other treatment. From Fig. 5.8c, it is clear that these films are almost colorless, whether the film is a single layer of 5 nm thick or 20 layers of 200 nm thick, indicating the good transparency of the film. When the applied voltage is 0 V, the transmittance of this tungsten oxide nanowire films lies in the wavelength range of 450–700 nm and changes from 95, 95, 90, 87, and 83 to 76% for monolayer, 2 layer, 5 layer, 10 layer, 15 layer, and 20 layer of $W_{18}O_{49}$ nanowire films, respectively. As a result, the ultrathin $W_{18}O_{49}$ nanowire electrode adjusted for a glass substrate exhibits great potential for use in high transparency electrochromic devices by enhancing the total optical transmittance in the visible region. When the applied voltage is −1 V, the color of the film rapidly changes to blue, and the color depth increases as the film thickness increases (Fig. 5.8d). Under different applied voltages +1, +0.3, −0, −0.3, −0.5, −0.7, and −1 V, the film color and transmittance have obvious change. The color of the film becomes darker as the applied voltage becomes smaller, but the color change is less obvious when the applied voltage is lower than −1.0 V [33]. When the thickness is increased to 20 layers, the film becomes dark blue. These apparent color changes on the surface of ordered tungsten oxide nanowire films have great potential for applications. Hence, the reduction in film thickness increases the rate of ion adsorption and desorption of EC materials, thereby increasing the sensitivity of the device. However, when the film thickness is only a single layer, its color change is very weak between 1 and −1 V. When a voltage of +1.0 V is applied to the dark blue film, the film quickly becomes fade and the transparent film is again obtained. Figure 5.8d, e shows the coloring and fading properties of the films, tested at 632.8 nm absorption wavelength. The time of coloring and fading increases gradually with the film thickness. The coloring/fading time of monolayer nanowire films can be as low as 2 s, which is faster than the titanium oxide nanostructures [34] and amorphous and crystalline WO_3 structures [35]. With the increase in the thickness, the coloring and fading time of the film gradually becomes longer. Moreover, when the thickness of the film approaches to 20 layers, the film coloring/fading time reaches 30 s.

Hence, the large active surface and ultrathin thickness of the as-assembled nanowire films are considered important factors for fast switching time, which makes the nanowire device appropriate for electrochromic devices. Furthermore, while making a comparison between the randomly and ordered arranged nanowires with the constant nanowire density, the electrochromic performance and the repeatability of the disordered nanowire films are appeared to be unstable and

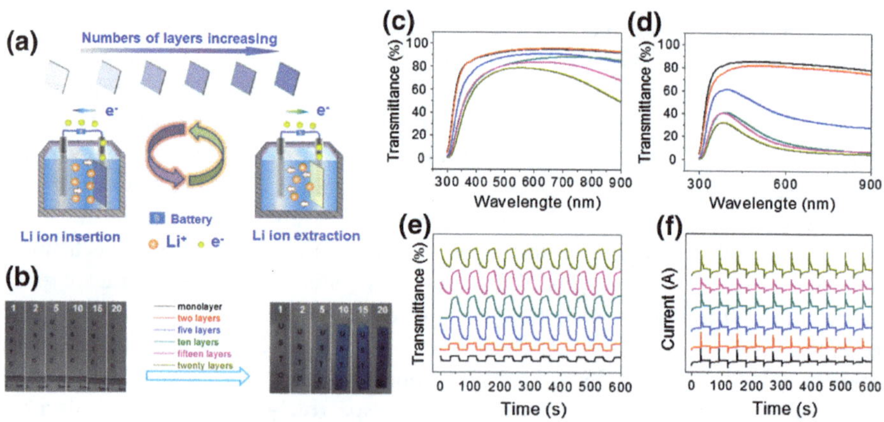

Fig. 5.8 **a** Demonstration of reversible electrochromic $W_{18}O_{49}$ nanowire film device. **b** Image of ordered $W_{18}O_{49}$ nanowire films with different thickness on ITO. **c**, **d** Optical transmittance of ordered $W_{18}O_{49}$ nanowire films with different thickness on ITO substrate. **e**, **f** Electrochromic switching of the ordered $W_{18}O_{49}$ nanowire films. Reprinted with the permission from Ref. [36]. Copyright 2013 American Chemical Society

poorer. The periodicity of the nanowires can be altered by controlling the surface pressure, and also, the electrochromic performance shows more improvement when the ordering degree increases.

5.7 Nanowire Films for Surface-Enhanced Raman Substrates

Based on capillary assembly, the integrated silver nanowire films can behave as an active platform for portable and recyclable SERS detection [35, 37]. The enhanced electric field of the noble metal nanostructures is the main influencing factor of SERS enhancement. Different silver nanostructures have been studied for SERS, such as nanowires, nanoparticles [38], nanosheets [39], and vertically grown nanorods [40–42]. The schematic design for the cyclic utilization of SERS is shown in Fig. 5.9a. The Raman peaks of Rhodamine 6G (R6G) appear at peaks 1310, 1362, 1460, 1508, and 1648 cm^{-1}, respectively (Fig. 5.9b), which are consistent with those reported in the literatures. Moreover, Fig. 5.9b shows an eight-cycle test showing the stability of the Raman substrate. The Ag-nanowire film formation in a confined capillary provides a unique structure that will open up potential applications in future; i.e., Raman enhancement of Ag-nanowire film by the capillary method is stronger than that of the conventional planar silver nanostructures obtained from LB as shown by the red line in Fig. 5.9b, whereas SERS spectrum of Ag colloid nanoparticle (highlighted by orange line) shows that films exhibit the worst performance (Fig. 5.9b).

The SERS performance of disordered Ag-nanowire SERS shows the unstable state as compared with ordered Ag-nanowire films by capillary or even with LB technique (Fig. 5.10a, b). In order to make the comparison, glass capillaries are fabricated with different bending radius using oxyhydrogen flame, exhibiting better SERS performance as the bending radius of the glass capillary increases (Fig. 5.10c). Moreover, the silver nanowire film on the inner wall of the capillary facilitates the portable and recycling of the SERS. Figure 5.9b shows that no Raman signal is observed by immersing the SERS using R6G as the probe molecule in concentrated hydrochloric acid liquid, showing that Ag-nanowire films in the capillary can be used as a reusable, portable, and durable substrate. Figure 5.9c indicates the reproducibility of SERS after eight rounds of recycling with the error bar also representing the stability of the substrate. Furthermore, Fig. 5.10c shows that there is no perturbation (black) in the test, which means there is no overlap of a high SERS state (black) and a low Raman signals (red) [16].

In addition, one of the most significant applications of the SERS is the detection of contaminants. However, accurate detection of the mixture is very important as contaminant samples are typically mixtures of multiple components. In order to further explore the application of silver nanowire thin film substrate, SERS test is further applied to the mixed solution of 4-MBA and R6G. A comparison of the

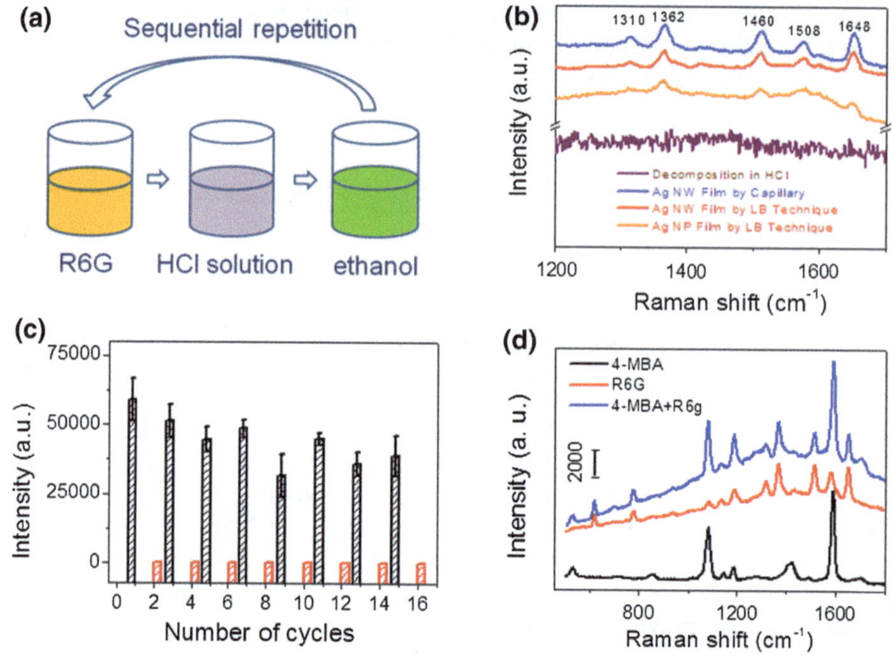

Fig. 5.9 **a** Schematic design of the cyclic utilization of SERS. **b** SERS spectra of adsorbed Rhodamine 6G (131,027 mol/L). **c** Repeatability of the reusable capillary SERS about 8 cycles. **d** SERS spectra of adsorbed R6G, 4-MBA, and a mixture of R6G and 4-MBA. Reprinted with the permission from Ref. [16]. Copyright 2012 Nature Publishing Group

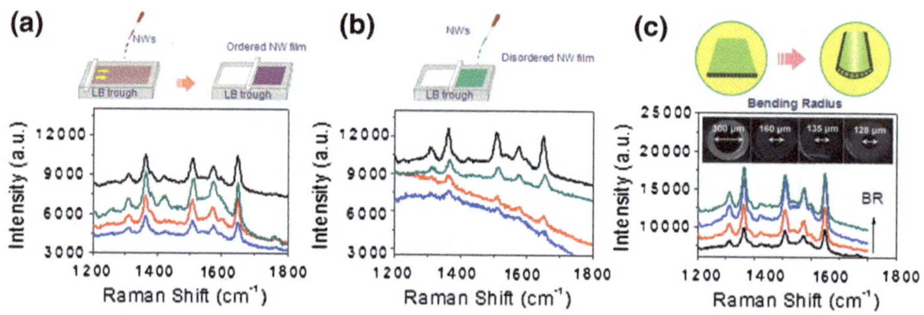

Fig. 5.10 | SERS performances of **a** ordered, **b** disordered, and **c** cambered Ag-nanowire films. Reprinted with the permission from Ref. [16]. Copyright 2012 Nature Publishing Group

three spectra in Fig. 5.10 demonstrates that the capillary SERS provides a versatile platform that can even detect mixture pollutions. Moreover, the order of the silver nanostructures has different enhancement effects in the SERS test. Hence, the SERS thin films prepared at different flow rates exhibit different SERS intensities, indicating that the SERS properties are strongly dependent on the density and ordering of the nanowires [16].

5.8 Flexible Transparent Electrodes Based on the Nanowire Assemblies

The effective design of high-quality flexible transparent electrodes is largely reliant on their electrical conductivity, transmittance, and the balance between their transmittance and conductivity. After customizing the transmittance of a single layer of the nanowire, more complex structures can be designed to study their conductivity. A process for manufacturing a flexible transparent electrode is shown in Fig. 5.11a [43]. A two-layer nanowire setup (Ag and Te) has a mesh-like structure formed by depositing the desired crossing angle with the LB technique that allows for flexible interpolation of transmittance while maintaining high-performance conductivity. The nanowires can be selectively oxidized and removed due to the different activity of the two kinds of nanowires. By this technique, highly flexible and transparent Ag-nanowire films on PET substrate can be manufactured on a large scale as shown in Fig. 5.11b. The Ag and Te nanowire co-assemblies are shown in Fig. 5.11c–f, showing Te nanowire assembly in between the Ag nanowires and providing space between Ag nanowires. The transmittance of the nanowire films is improved by loading the other nanowire monolayer on the first one while removing the Te nanowires (Fig. 5.11e). The assembled Ag-nanowire films with controlled nanowire spacing show optically transparency and mechanically flexibility [43].

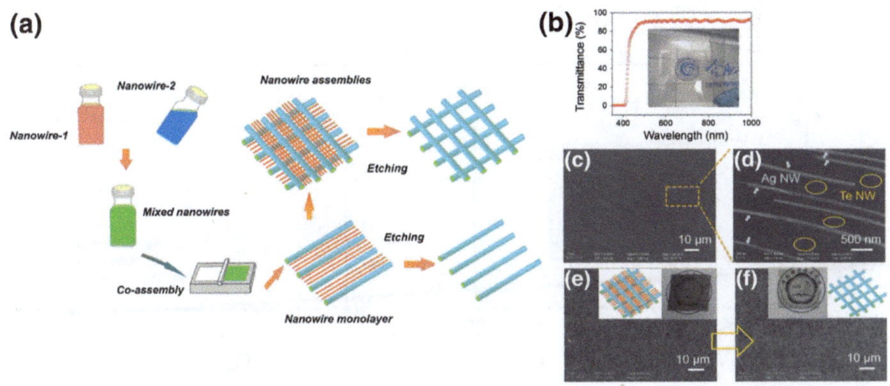

Fig. 5.11 **a** Schematic design of nanowire assembly to make flexible transparent electrodes. **b** The optical transmittance spectra of the nanowire-based transparent electrode. **c**, **d** SEM images of nanowire co-assemblies and monolayers of Ag and Te nanowires at different magnifications. **e**, **f** SEM images of the nanowire-assembled networks before and after etching away of the Te nanowires. Reprinted with the permission from Ref. [43]. Copyright 2014 WILEY-VCH

Moreover, by using Te nanowires, the pitch of well-aligned Ag nanowires can be regulated. The film formed by regulating the molar ratio of Ag nanowire and Te nanowire from 1:400, 1:350, 1:300, 1:250, 1:200, and 1:100 is shown in Fig. 5.12b–g, showing the decrease in the pore area of the Ag-nanowire assemblies as the amount of Te nanowires decreases in the co-assembly process. So, the pore area of the Ag-nanowire films ranges from hundreds of square micron (Fig. 5.12b) to hundreds of square nanometers (Fig. 5.12g), resulting in the change the transmission from 97.3 to 92, 90, 85, 80, and 73.8%, respectively (Fig. 5.12h, i) [43].

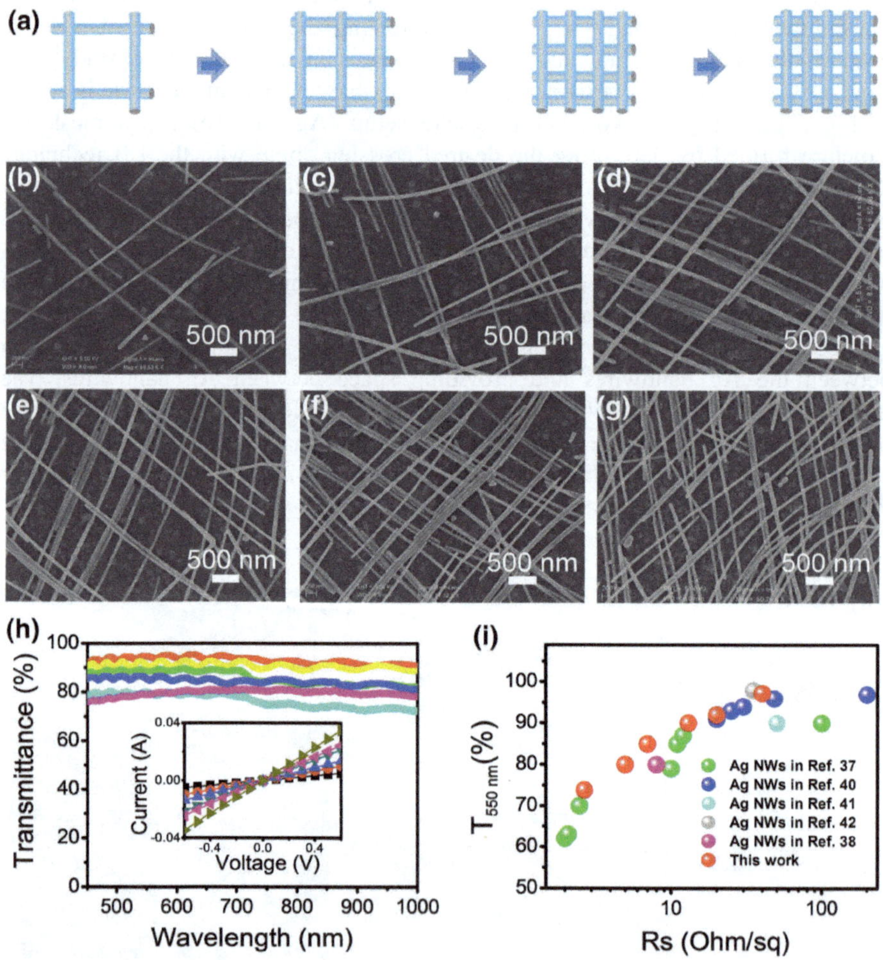

Fig. 5.12 **a** Design of the pore area of the nanowire networks. **b**–**g** SEM images of well-defined Ag-nanowire networks with different pore areas of the nanowire assemblies. **h** The optical transmittance spectra corresponding to the samples from (**b**) to (**g**). **i** Sheet resistance versus optical transmission (at 550 nm) for Ag-nanowire networks prepared by different methods. Reprinted with the permission from Ref. [43]. Copyright 2014 WILEY-VCH

References

1. Chen K-I, Li B-R, Chen Y-T (2011) Silicon nanowire field-effect transistor-based biosensors for biomedical diagnosis and cellular recording investigation. Nano Today 6(2):131–154
2. Yu G, Lieber CM (2010) Assembly and integration of semiconductor nanowires for functional nanosystems. Pure Appl Chem 82(12):2295–2314
3. Rueckes T, Kim K, Joselevich E, Tseng GY, Cheung C-L, Lieber CM (2000) Carbon nanotube-based nonvolatile random access memory for molecular computing. Science 289 (5476):94–97
4. Das S, Gates AJ, Abdu HA, Rose GS, Picconatto CA, Ellenbogen JC (2007) Designs for ultra-tiny, special-purpose nanoelectronic circuits. IEEE Trans Circuits Syst I Regul Pap 54 (11):2528–2540
5. Cui Y, Lieber CM (2001) Functional nanoscale electronic devices assembled using silicon nanowire building blocks. Science 291(5505):851–853
6. Huang Y, Duan X, Cui Y, Lauhon LJ, Kim K-H, Lieber CM (2001) Logic gates and computation from assembled nanowire building blocks. Science 294(5545):1313–1317
7. Zhong Z, Wang D, Cui Y, Bockrath MW, Lieber CM (2003) Nanowire crossbar arrays as address decoders for integrated nanosystems. Science 302(5649):1377–1379
8. Yan H, Choe HS, Nam S, Hu Y, Das S, Klemic JF, Ellenbogen JC, Lieber CM (2011) Programmable nanowire circuits for nanoprocessors. Nature 470(7333):240–244
9. Dong Y, Yu G, McAlpine MC, Lu W, Lieber CM (2008) Si/a-Si core/shell nanowires as nonvolatile crossbar switches. Nano Lett 8(2):386–391
10. McAlpine MC, Ahmad H, Wang D, Heath JR (2007) Highly ordered nanowire arrays on plastic substrates for ultrasensitive flexible chemical sensors. Nat. Mat. 6(5):379–384
11. Shimizu KT, Fabbri JD, Jelincic JJ, Melosh NA (2006) Soft deposition of large-area metal contacts for molecular electronics. Adv Mat 18(12):1499–1504
12. Zeng J, Huang JL, Liu C, Wu CH, Lin Y, Wang XP, Zhang SY, Hou JG, Xia YN (2010) Gold-based hybrid nanocrystals through heterogeneous nucleation and growth. Adv Mater 22 (17):1936
13. Wang L, Chen D (2006) A one-pot approach to the preparation of silver-PMMA "shell-core" nanocomposite. Colloid Polym Sci 284(4):449–454
14. Hu B, Yu SH, Wang K, Liu L, Xu XW (2008) Functional carbonaceous materials from hydrothermal carbonization of biomass: an effective chemical process. Dalton Trans 40:5414–5423
15. Gordon R, Sinton D, Kavanagh KL, Brolo AG (2008) A new generation of sensors based on extraordinary optical transmission. Acc Chem Res 41(8):1049–1057
16. Liu J-W, Wang J-L, Huang W-R, Yu L, Ren X-F, Wen W-C, Yu S-H (2012) Ordering Ag nanowire arrays by a glass capillary: a portable, reusable and durable SERS substrate. Sci Rep 2
17. Bartlett RS (1925) Photo-resistance effect for metals at low temperatures. Phy Rev 26(2):247
18. Liu J-W, Zhu J-H, Zhang C-L, Liang H-W, Yu S-H (2010) Mesostructured assemblies of ultrathin superlong tellurium nanowires and their photoconductivity. J Am Chem Soc 132 (26):8945–8952
19. Panahi-Kalamuei M, Mohandes F, Mousavi-Kamazani M, Salavati-Niasari M, Fereshteh Z, Fathi M (2014) Tellurium nanostructures: Simple chemical reduction synthesis, characterization and photovoltaic measurements. Mater Sci Semicond Process 27:1028–1035
20. Loferski JJ (1954) Infrared optical properties of single crystals of tellurium. Phy Rev 93 (4):707
21. Liu JW, Xu J, Liang HW, Wang K, Yu SH (2012) Macroscale ordered ultrathin telluride nanowire films, and tellurium/telluride hetero-nanowire films. Angw Chem In Ed 51 (30):7420–7425
22. Xi G, Ouyang S, Li P, Ye J, Ma Q, Su N, Bai H, Wang C (2012) Ultrathin W18O49 nanowires with diameters below 1 nm: synthesis, near-infrared absorption,

photoluminescence, and photochemical reduction of carbon dioxide. Angew Chem In Ed 51 (10):2395–2399
23. Di Ventra M, Pershin YV, Chua LO (2009) Circuit elements with memory: memristors, memcapacitors, and meminductors. IEEE Trans Circ Theory 97(10):1717–1724
24. Liang L, Zhang J, Zhou Y, Xie J, Zhang X, Guan M, Pan B, Xie Y (2013) High-performance flexible electrochromic device based on facile semiconductor-to-metal transition realized by WO_3 [middot] $2H_2O$ ultrathin nanosheets. Sci Rep 3:1936
25. Granqvist C-G, Lansåker P, Mlyuka N, Niklasson G, Avendano E (2009) Progress in chromogenics: new results for electrochromic and thermochromic materials and devices. Sol Energy Mater Sol Cells 93(12):2032–2039
26. Deb SK (2008) Opportunities and challenges in science and technology of WO 3 for electrochromic and related applications. Sol Energy Mater Sol Cells 92(2):245–258
27. Yoo SJ, Lim JW, Sung Y-E, Jung YH, Choi HG, Kim DK (2007) Fast switchable electrochromic properties of tungsten oxide nanowire bundles. Appl Phys Lett
28. Yao C-J, Zhong Y-W, Nie H-J, Abruña HCD, Yao J (2011) Near-IR electrochromism in electropolymerized films of a biscyclometalated ruthenium complex bridged by 1, 2, 4, 5-tetra (2-pyridyl) benzene. J Am Chem Soc **133**(51):20720–20723
29. Granqvist C-G (2006) Electrochromic materials: out of a niche. Nat Mat 5(2):89–90
30. Deb S (1969) A novel electrophotographic system. App Opt 8(101):192–195
31. Niklasson GA, Granqvist CG (2007) Electrochromics for smart windows: thin films of tungsten oxide and nickel oxide, and devices based on these. J Mat Chem 17(2):127–156
32. Granqvist CG (2014) Electrochromics for smart windows: oxide-based thin films and devices. Thin Solid Films 564:1–38
33. Wang J, Khoo E, Lee PS, Ma J (2008) Synthesis, assembly, and electrochromic properties of uniform crystalline WO3 nanorods. J Phy Chem C 112(37):14306–14312
34. Chen J-Z, Ko W-Y, Yen Y-C, Chen P-H, Lin K-J (2012) Hydrothermally processed TiO2 nanowire electrodes with antireflective and electrochromic properties. ACS Nano 6(8):6633–6639
35. Shi HY, Hu B, Yu XC, Zhao RL, Ren XF, Liu SL, Liu JW, Feng M, Xu AW, Yu SH (2010) Ordering of disordered nanowires: spontaneous formation of highly aligned, ultralong Ag nanowire films at oil–water–air interface. Adv Mat 20(6):958–964
36. Liu BJ-W, Zheng J, Wang J-L, Xu J, Li H-H, Yu S-H (2013) Ultrathin W18O49 nanowire assemblies for electrochromic devices. Nano Lett 13(8):3589–3593
37. Tao AR, Yang P (2005) Polarized surface-enhanced Raman spectroscopy on coupled metallic nanowires. J Phy Chem B 109(33):15687–15690
38. Yang C, Gu H, Lin W, Yuen MM, Wong CP, Xiong M, Gao B (2011) Silver nanowires: from scalable synthesis to recyclable foldable electronics. Adv Mat 23(27):3052–3056
39. Liu H, Lalanne P (2008) Microscopic theory of the extraordinary optical transmission. Nature 452(7188):728–731
40. Qian H-S, Yu S-H, Gong J-Y, Luo L-B, Fei L-F (2006) High-quality luminescent tellurium nanowires of several nanometers in diameter and high aspect ratio synthesized by a poly (vinyl pyrrolidone)-assisted hydrothermal process. Langmuir 22(8):3830–3835
41. Zhang G, Yu Q, Yao Z, Li X (2009) Large scale highly crystalline Bi2Te3 nanotubes through solution phase nanoscale Kirkendall effect fabrication. Chem Comm 17:2317–2319
42. Gordon R, Sinton D, Kavanagh KL, Brolo AG (2008) A new generation of sensors based on extraordinary optical transmission. Acc Chem Res 41(8):1049–1057
43. Liu JW, Wang JL, Wang ZH, Huang WR, Yu SH (2014) Manipulating nanowire assembly for flexible transparent electrodes. Angew Chem Int Ed 53(49):13477–13482

FSC
www.fsc.org
MIX
Papier | Fördert gute Waldnutzung
FSC® C083411

Kolibri 360 GmbH
Ferdinand-Jühlke-Straße 7
99095 Erfurt, Deutschland
produktsicherheit@kolibri360.de